AQA Mathematics

for GCSE

Exclusively endorsed and approved by AQA

Series Editor
Paul Metcalf
Series Advisor
David Hodgson
Lead Author
Margaret Thornton

June Haighton
Anne Haworth
Janice Johns
Steven Lomax
Andrew Manning
Kathryn Scott
Chris Sherrington
Mark Willis

FOUNDATION
Linear 1

Published in 2006 by:
Nelson Thornes Ltd
Delta Place
27 Bath Road
CHELTENHAM
GL53 7TH
United Kingdom

07 08 09 10 / 10 9 8 7 6 5 4 3

A catalogue record for this book is available from the British Library.

ISBN 13: 978 0 7487 9750 9

Cover photograph: Salmon by Kyle Krause/Index Stock/OSF/Photolibrary
Illustrations by Roger Penwill
Page make-up by MCS Publishing Services Ltd, Salisbury, Wiltshire

Printed in Great Britain by Scotprint

Acknowledgements

The authors and publishers wish to thank the following for their contribution:
David Bowles for providing the Assess questions
David Hodgson for reviewing draft manuscripts

Thank you to the following schools:
Little Heath School, Reading
The Kingswinford School, Dudley
Thorne Grammar School, Doncaster

The publishers thank the following for permission to reproduce copyright material:

Explore photos
Diver – Corel 55 (NT); Astronaut – Digital Vision 6 (NT);
Mountain climber – Digital Vision XA (NT); Desert explorer – Martin Harvey/Alamy.

Footballers – Corel 778 (NT); Archery – AAS; Plant – Stockbyte 29 (NT);
Compass – Stockbyte 35 (NT); British banknotes – Corel 590 (NT); Lighthouse – Corel 502 (NT);
Rows of Colorful Seats in Empty Stadium – Paul Gun/CORBIS;
Alnwick Castle – Malcom Fife/zefa/Corbis; US Dollar bill – Photodisc 68 (NT);
Rabbit G/V Hart/Photodisc 50 (NT); Discs – Nick Koudis/Photodisc 37 (NT);
Coins – Corel 633 (NT); Students in exam – Digital Stock 10 (NT); Beach – Corel 777 (NT);
Goal kick – David Madison/CORBIS; Vitruvian man – Corel 481 (NT); Coins – Corel 590 (NT);
Electricity pylons – Photodisc 4 (NT); Dinner ladies – Shout/Alamy;
Lydia & George – Jacqueline Belanger; Crufts dog show – Homer Sykes/Alamy;
Shopping centre – Corel 641 (NT); Telephone survey – Photodisc 55 (NT);
Discs – Nick Koudis/Photodisc 37 (NT).

The publishers have made every effort to contact copyright holders but apologise if any have been overlooked.

Contents

Introduction

This book has been written by teachers and examiners who not only want you to get the best grade you can in your GCSE exam but also to enjoy maths.

Each chapter has the following stages:

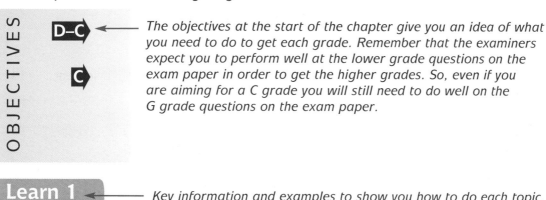

OBJECTIVES

The objectives at the start of the chapter give you an idea of what you need to do to get each grade. Remember that the examiners expect you to perform well at the lower grade questions on the exam paper in order to get the higher grades. So, even if you are aiming for a C grade you will still need to do well on the G grade questions on the exam paper.

Learn 1

Key information and examples to show you how to do each topic. There are several Learn sections in each chapter.

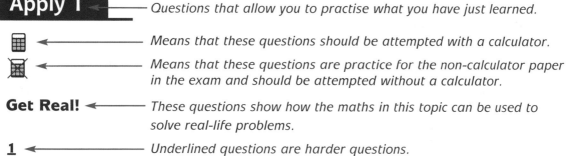

Apply 1

Questions that allow you to practise what you have just learned.

Means that these questions should be attempted with a calculator.

Means that these questions are practice for the non-calculator paper in the exam and should be attempted without a calculator.

Get Real!

These questions show how the maths in this topic can be used to solve real-life problems.

1

Underlined questions are harder questions.

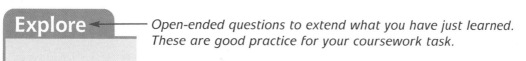

Explore

Open-ended questions to extend what you have just learned. These are good practice for your coursework task.

ASSESS

End of chapter questions written by an examiner.

Some chapters feature additional questions taken from real past papers to further your understanding.

1 Statistical measures

OBJECTIVES

G **Examiners would normally expect students who get a G grade to be able to:**

Find the mode for a set of numbers

Find the median for an odd set of numbers

F **Examiners would normally expect students who get an F grade also to be able to:**

Work out the range for a set of numbers or for a graph

Calculate the mean for a set of numbers

Find the median for an even set of numbers

Write down the mode from a graph

E **Examiners would normally expect students who get an E grade also to be able to:**

Calculate the '*fx*' column for a frequency distribution

Compare the mean and range of two distributions

D **Examiners would normally expect students who get a D grade also to be able to:**

Calculate the mean for a frequency distribution

Find the modal class for grouped data

C **Examiners would normally expect students who get a C grade also to be able to:**

Find the mean for grouped data

Find the median class for grouped data

What you should already know ...

■ Understand addition, multiplication and division of a set of numbers – with and without the use of a calculator

VOCABULARY

Average – a single value that is used to represent a set of data

Mode – the value that occurs most often

Modal class – the class with the highest frequency

Median – the middle value when all the values have been arranged in order of size; for an even set of numbers, the median is the mean of the two middle values

Mean – found by calculating $\dfrac{\text{the total of all the values}}{\text{the number of values}}$

Range – a measure of spread found by calculating the difference between the largest and smallest values in the data, for example, the range of 1, 2, 3, 4, 5 is $5 - 1 = 4$

Frequency table or **frequency distribution** – a table showing how frequently each quantity occurs, for example,

Number in family	2	3	4	5	6	7	8
Frequency	2	3	8	4	2	0	1

Data – information that has been collected

Discrete data – data that can only be counted and take certain values, for example, numbers of cars (you can have 3 cars or 4 cars but nothing in between, so $3\frac{1}{2}$ cars is not possible)

Continuous data – data that can be measured and take any value; length, weight and temperature are all examples of continuous data

Grouped data – data that has been grouped into specific intervals

Learn 1   Three averages and a range

These are all 'averages'

Example:

From this set of data find the mean , mode , median and range of the data.

2 1 2 3 5 5 1 5 12

The range is the difference between the largest and smallest numbers. It is a measure of spread

The mean $= \dfrac{\text{The total of all the values}}{\text{The total number of values}}$

$= \dfrac{2 + 1 + 2 + 3 + 5 + 5 + 1 + 5 + 12}{9} = \dfrac{36}{9} = 4$

The mode is the value that occurs most often.

2 1 2 3 5 5 1 5 12

There are three fives. Five is the **most** popular so the mode = 5

The median is the middle value when all the values have been arranged **in order of size.**

Arranging in order of size:

1 1 2 2 3 5 5 5 12

The **middle** number is 3 so the median = 3

If there are two middle numbers then find the mean of the two numbers

Range   Largest number = 12

Smallest number = 1

The range is the difference between the largest and smallest numbers = $12 - 1 = 11$

Apply 1

 1 Find   **a** the median   **b** the mode   of the following sets of data.

   i   4   8   3   7   6   8   3   3   2   4   3   5   2   3   6

   ii   1   3   4   2   2   3   4   5   8   6   5   4   5   8   9

   iii 2   4   3   6   5   7   7   8   4   8   2   4   6   7   7

 2 Find   **a** the mean   **b** the range   of the following sets of data.

   i   3   2   3   2   5

   ii   1   4   4   5   1

   iii 2   3   1   4   5   6   7   4   2   3

  3 Find   **a** the median   **b** the mode   **c** the mean   **d** the range
of the following sets of data.

   i 16   19   14   15   15   11

   ii 18   26   23   23   37   23

 <u>**4**</u> Find the mode, the median and the mean of these sets of data.

   a A dice is rolled nine times and these scores are recorded.
     4   1   1   4   2   3   2   6   4

   b A group of eleven fathers were asked how many children they had.
     4   3   1   2   1   1   1   3   2   3   1

   c A local football team plays ten matches and lets in these number of goals.
     1   0   2   4   0   3   2   1   4   1

   d A shopkeeper keeps a record of the number of broken eggs found in
eight deliveries.
     5   1   5   1   2   0   5   2

  <u>**5**</u> The mean of three numbers is 5. Two of the numbers are 3 and 4.
What is the third number?

 6 Get Real!
Zoe was having a good year scoring goals in the U16 hockey team.
After 10 matches she had scored: 1, 2, 2, 2, 0, 4, 3, 3, 1, 2 goals.
She said 'A mean of 2 goals is not bad'.
Check to see if she was right in thinking that the mean was 2 goals.

  7 Get Real!
In a football tournament, the team from Eastgate School
scored 2, 3, 3, 3, 3, 4 goals.
The team from Westgate School scored 1, 2, 2, 3, 5, 5 goals.
By working out and comparing the mean and range, which team
do you think played better in the tournament?

8 Get Real!

Mr Hobbs, the mathematics teacher, decided to give his class a test each day for two weeks.

a Tariq results were 4, 2, 4, 5, 5, 9, 6, 6, 8, 1.
Calculate the mean and range of his marks.

b Jennifer's results were 4, 4, 7, 5, 5, 7, 6, 4, 3, 5.
Calculate the mean and range of her marks.

c Who do you think did better? Give a reason for your answer.

9 Get Real!

Emily, a botanist, is testing various seeds. She measures the heights of four different plants after two weeks of growth.
She draws a bar chart of her results.

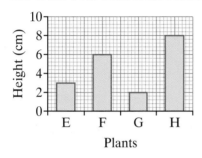

a Calculate the mean height of the plants.

b What is the range of heights?

10 Get Real!

The heights of five men are 177.8 cm, 175.5 cm, 174 cm, 179.9 cm and 176.2 cm.
What is the mean height of the 5 men?

11 Get Real!

There are 6 people in a lift. Their weights, to the nearest kilogram, are 84, 73, 80, 67, 82 and 76 kg.
Find

a their median weight

b their mean weight.

Learn 2   Mean, mode, median and range for a frequency table

Example: From this table work out the mean, mode, median and range of the number of goals scored.

Goals scored (x)	Frequency (f)
0	3
1	6
2	5
3	3
4	2
5	1

Completing the table:

Goals scored (x)	Frequency (f)	Frequency × goals scored (fx)
0	3	$0 \times 3 = 0$
1	6	$1 \times 6 = 6$
2	5	$2 \times 5 = 10$
3	3	$3 \times 3 = 9$
4	2	$4 \times 2 = 8$
5	1	$5 \times 1 = 5$
Total	$\Sigma f = 3+6+5+3+2+1 = 20$	$\Sigma fx = 0+6+10+9+8+5 = 38$

Σ means 'the sum of' so Σf means the sum of the frequencies and Σfx means the sum of the (frequency × goals scored)

The mean $= \dfrac{\Sigma fx}{\Sigma f} = \dfrac{38}{20} = 1.9$ goals

The mode is the value that occurs the most often.

Mode = 1 goal   In a frequency table, this is the value with the highest frequency. The highest frequency is 6 so the mode is 1

The median is the middle value when all the values have been arranged **in order of size**.

The middle value is the $\dfrac{\Sigma f + 1}{2}$ th value $= \dfrac{20 + 1}{2}$ th value $= 10.5$th value.

So the median lies between the 10th and 11th value.

To find the 10th and 11th terms do a quick running total of the frequencies.

Goals scored (x)	Frequency (f)	Running total
0	3	3
1	6	$3 + 6 = 9$
2	5	$9 + 5 = 14$
3	3	
4	2	
5	1	

Range   Largest number $= 5$ The 10th and 11th terms are in this class
         Smallest number $= 0$ so the median = 2 goals

The range is the difference between the largest and smallest numbers $= 5 - 0 = 5$

Apply 2

 1 The shoe sizes of ten people are shown in the table.

Shoe size (x)	Frequency (f)	Frequency × shoe size (fx)
3	3	
4	3	
5	4	

a Copy and complete the table.   **b** Find the mean shoe size.

2 A dice is thrown 100 times. The scores are shown in the table.

Score (x)	Frequency (f)	Frequency × score (fx)
1	18	
2	19	
3	16	
4	12	
5	15	
6	20	

Copy and complete the table and find the mean score.

3 Ten people were asked to give the ages of their cars.
Their answers are shown in the table.

Age of car (x)	Frequency (f)
1	2
2	3
3	4
4	1

Tom says that the mean age of the cars is 6 years.

a Find the mean.

b What do you think Tom did wrong?

4 Get Real!

Andrew has five people in his family. He wondered how many people
there were in his friends' families. He asked 20 of his friends and put his
results in a table.

Number in family	2	3	4	5	6	7	8
Frequency	2	3	8	4	2	0	1

From the data calculate:

a the median   **b** the mode   **c** the mean   **d** the range.

5 Get Real!

A group of 30 teenagers were learning archery
and were allowed ten shots each at the target.
The instructor counted the number of times they hit
the target and recorded the following results.

2 6 2 0 2 3 8 7 2 5 6 1 7 4 8
8 6 1 0 3 2 2 1 3 8 3 2 6 7 5

a Copy this table and use the figures to complete it.

Number of hits	0	1	2	3	4	5	6	7	8
Frequency									

b Use the table to calculate:

i the mode   **ii** the median   **iii** the mean   **iv** the range.

6 Get Real!

Rachel is investigating the number of letters in words in her reading book.
Her results are shown in the following table.

Number of letters in a word	Frequency
1	5
2	6
3	10
4	4
5	7
6	8
7	4
8	2

She says, 'The mode is 10, the median is 4 and the mean is 4.2'.
Is she right?
Give a reason for your answer.
Which average do you think best represents the data? Why?

7 Get Real!

Mr Farrington, a technology teacher, has estimated that the average length of a box of bolts is 60 mm, to the nearest millimetre. The class measured the lengths of 20 boxes and recorded their results in a table.

Length (mm)	Frequency	Frequency × length
58	3	
59	4	
60	2	
61	2	
62	8	
63	1	
Total		

Copy and complete the table. Work out the mean length.
Was Mr Farrington correct in his estimation?

8 Get Real!

These marks were obtained by a class of 28 students in a science test.
The maximum mark possible was 25.

20 17 21 15 16 15 12 14 15 19 21 17 20 22
16 19 21 13 22 15 14 18 20 13 18 16 15 19

By drawing a frequency table:

a Calculate the mean and work out the median and mode.

b Work out the range of the marks.

c What would you give as the pass mark? Explain your answer.

Explore

A DIY shop sells a selection of large letters that can be used to design name boards for houses

The shopkeeper wants to buy 1000 letters but realises that some letters will be more popular than other letters

For example

ROSE COTTAGE

needs two Es, two Os, two Ts and one of each of the other letters

How many of each letter should the shopkeeper buy?

Investigate further

Explore

- ◎ Draw a straight line on a sheet of paper
- ◎ Place a ruler some distance away from the line
- ◎ Let several people look at the line and the ruler
- ◎ Ask each one to estimate the length of the line
- ◎ Put your results into a frequency table and calculate the mean
- ◎ How does this mean compare with the actual length of the line?

Investigate further

Learn 3   Mean, mode and median for a grouped frequency table

Example:

The weights of 50 potatoes are measured to the nearest gram and shown in the table below.
From this table work out the mean, modal class and the class containing the median.

Weight in grams	Frequency
75–79	3
80–84	3
85–89	3
90–94	10
95–99	7
100–104	7
105–109	5
110–114	4
115–119	2
120–124	4
125–129	1
130–134	1

The weight of potatoes is continuous

The midpoint of the class is
$$\frac{90 + 94}{2} = 92$$

Completing the table:

Weight in grams	Frequency (f)	Midpoint (x)	Frequency × midpoint (fx)
75–79	3	77	77 × 3 = 231
80–84	3	82	82 × 3 = 246
85–89	3	87	87 × 3 = 261
90–94	10	92	92 × 10 = 920
95–99	7	97	97 × 7 = 679
100–104	7	102	102 × 7 = 714
105–109	5	107	107 × 5 = 535
110–114	4	112	112 × 4 = 448
115–119	2	117	117 × 2 = 234
120–124	4	122	122 × 4 = 488
125–129	1	127	127 × 1 = 127
130–134	1	132	132 × 1 = 132
Totals	$\sum f = 50$		$\sum fx = 5015$

Start by finding the midpoint of each class and then continue as a frequency table

$\sum fx = 231 + 246 + 261 + 920 + 679 + 714 + 535 + 448 + 234 + 488 + 127 + 132$

The mean $= \dfrac{\sum fx}{\sum f} = \dfrac{5015}{50} = 100.3$ g

Remember that this is only an estimate as you have used the midpoints

The modal class is the class that occurs the most often.

Modal class = 90–94 g

In a frequency table, this is the class with the highest frequency

The median is the middle value when all the values have been arranged **in order of size**.

The middle value is the $\dfrac{\sum f + 1}{2}$ th value $= \dfrac{50 + 1}{2}$ th value = 25.5th value.

So the median lies between the 25th and 26th value.

To find the 25th and 26th terms do a quick running total of the frequencies.

Weight in grams	Frequency	Running total
75–79	3	3
80–84	3	3 + 3 = 6
85–89	3	6 + 3 = 9
90–94	10	9 + 10 = 19
95–99	7	19 + 7 = 26
100–104	7	
105–109	5	
110–114	4	
115–119	2	
120–124	4	
125–129	1	
130–134	1	

The 25th and 26th values are in this class so the class containing the median = 95–99 g

Apply 3

1 Get Real!

The table shows the wages of 40 staff in a small company.

Wages (£)	Frequency
$50 \leqslant x < 100$	5
$100 \leqslant x < 150$	13
$150 \leqslant x < 200$	11
$200 \leqslant x < 250$	9
$250 \leqslant x < 300$	0
$300 \leqslant x < 350$	2

Find:

a the modal class

b the class that contains the median

c an estimate of the mean.

2 Get Real!

The scores obtained in a survey of reading ability are given in this table.

Reading scores (x)	Frequency
$0 \leqslant x < 5$	15
$5 \leqslant x < 10$	60
$10 \leqslant x < 15$	125
$15 \leqslant x < 20$	260
$20 \leqslant x < 25$	250
$25 \leqslant x < 30$	200
$30 \leqslant x < 35$	90

a What is the modal class?

b Calculate an estimated mean reading score.

3 Get Real!

The lengths, to the nearest millimetre, of a sample of a certain type of plant are given below.

51 51 58 54 59 60 52 52 55 49 51 53
55 60 58 57 51 57 56 50 53 58 59 57

a Calculate the mean length.

b Calculate an estimate of the mean length by grouping the data in class intervals of 47–49, 50–52, etc.

c Comment on your findings. What do you notice?

4 For each of these sets of data, work out the

   a mean   **b** modal class   **c** class containing the median group.

i

Mark	Frequency
21 up to 31	1
31 up to 41	1
41 up to 51	3
51 up to 61	9
61 up to 71	8
71 up to 81	6
81 up to 91	2

ii

Daily takings ($)	Frequency
480–499	2
500–519	3
520–539	5
540–559	7
560–579	11
580–599	13
600–619	6
620–639	4
640–659	0
660–679	1

5 The table shows the weights of 10 letters.

Weight (grams)	$0 \leqslant x < 20$	$20 \leqslant x < 40$	$40 \leqslant x < 60$	$60 \leqslant x < 80$	$80 \leqslant x < 100$
Number of letters	2	3	2	2	1

Calculate an estimate of the mean weight of a letter.

6 A survey was made of the amount of money spent at a supermarket by 20 shoppers. The table shows the results.

Amount spent, A (£)	$0 \leqslant A < 20$	$20 \leqslant A < 40$	$40 \leqslant A < 60$	$60 \leqslant A < 80$
Number of shoppers	1	7	8	4

Calculate an estimate of the mean amount of money spent by these shoppers.

Explore

  ◎ Sarah has collected some data from students

  ◎ She has found that the 'average' student is male, has brown eyes and hair, and is 165 cm tall

  ◎ Is this true for the students in your class?

  ◎ What else can you say about students in your class?

        Investigate further

Statistical measures

The following exercise tests your understanding of this chapter, with the questions appearing in order of increasing difficulty.

1 Find the mode, median, mean and range of the following sets of data.

a 4, 4, 2, 2, 2, 7, 7, 7, 1, 8, 7, 3, 3, 6

b 3, −3, −3, 2, −2, 1, −1, −1, 0, 0, 1, −1, 2, −1, −2, 3

2 Mr Chips records the marks of 10 students in his record book.

28   27   32   17   23   28   29   20   27   29

a Calculate:

i the mean mark

ii the median mark

iii the modal mark

iv the range.

He realises that the mark recorded as 32 should have been 35.

b What effect will this have on:

i the mean mark

ii the median mark

iii the modal mark

iv the range?

3 The table shows the number of trainers sold in one day in a sports shop.

Size	5	$5\frac{1}{2}$	6	$6\frac{1}{2}$	7
Frequency	10	15	9	3	10

Find the mode, median, mean and range of this data.

4 The information below shows the speeds of 60 white vans passing a speed camera. Find the class intervals that contain the mode and median and calculate an estimate of the mean.

Speed, x (mph)	$30 \leqslant x < 40$	$40 \leqslant x < 50$	$50 \leqslant x < 60$	$60 \leqslant x < 70$	$70 \leqslant x < 80$	$80 \leqslant x < 90$	$90 \leqslant x < 100$
Frequency	2	10	18	16	11	2	1

5 An office manager monitored the time members of staff took on 'private' telephone calls during working hours. Calculate an estimate of the mean length of the telephone calls, giving your answer to the nearest minute.

Time (nearest min)	3–5	6–8	9–12	13–16	17–20	21–25	26–30	31–40
Frequency	67	43	28	13	7	4	3	2

OBJECTIVES

Examiners would normally expect students who get an F grade to be able to:

Recognise acute, obtuse, reflex and right angles

Estimate angles and measure them accurately

Use properties of angles at a point and angles on a straight line

Understand the terms 'perpendicular lines' and 'parallel lines'

Examiners would normally expect students who get a D grade also to be able to:

Recognise corresponding angles and alternate angles

Understand and use three-figure bearings

What you should already know ...

■ Addition and subtraction of whole numbers

■ Simple fractions such as quarter, half, third, etc.

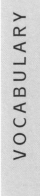

VOCABULARY

Revolution – one revolution is the same as a full turn or 360°

Right angle – an angle of 90°

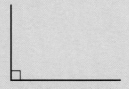

Acute angle – an angle between 0° and 90°

Obtuse angle – an angle greater than 90° but less than 180°

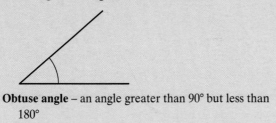

Reflex angle – an angle greater than 180° but less than 360°

Perpendicular lines – two lines at right angles to each other

Parallel lines – two lines that never meet and are always the same distance apart

Transversal – a line drawn across parallel lines

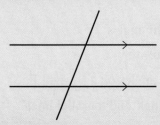

Alternate angles – the angles marked *a*, which appear on opposite sides of the transversal

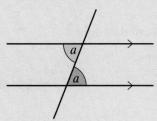

Corresponding angles – the angles marked *c*, which appear on the same side of the transversal

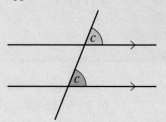

Bearing – an angle measured clockwise from North; all bearings should be written as three figure numbers, for example, 125° or 045°

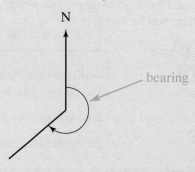

bearing

Learn 1   Full turns and part turns

Example:

What angle does the minute hand of a clock move through between 12:00 and 12:30?

The minute hand travels through half of a turn.

Half a turn $= \frac{1}{2} \times 360° = 180°$

There are 360° in a full turn

Apply 1

1 How many degrees are there in:

   a a quarter turn   **b** one third of a turn   **c** one twelfth of a turn   **d** five twelfths of a turn?

2 What fraction of a turn is:

    a 60°   **b** 45°   **c** 270°   **d** 300°?

3 **a** Jack faces north and makes a quarter turn clockwise.
Which way is he facing now?

    b Jill faces east and makes a half turn.
Which way is she facing now?

    c Huw faces west and makes a quarter turn anticlockwise.
Which way is he facing now?

    d Emma has made a half turn and is now facing east.
Where was she facing before she turned?

    e Ahmed has made a quarter turn clockwise and is now facing west.
Where was he facing before he turned?

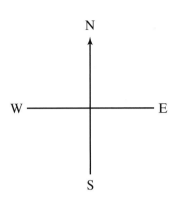

4 Get Real!

    a What angle does the **minute** hand of a clock move through between 1200 and 1220?

    b What angle does the **hour** hand of a clock move through between 0700 and 1400?

    c What is the angle between the hour hand and the minute hand at half past one?

Learn 2   Types of angles

Examples: Describe these angles:

a

An angle of 90° is called
a **right angle**.

It is marked with ⌐ as
shown above

c

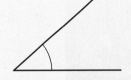

An angle less than 90° is
called an **acute angle**.

b

An angle between 90° and 180° is
called an **obtuse angle**.

d

An angle greater than 180°
but less than 360° is called
a **reflex angle**.

Apply 2

1 For each marked angle, write down whether it is an acute angle, an obtuse
   angle or a reflex angle.

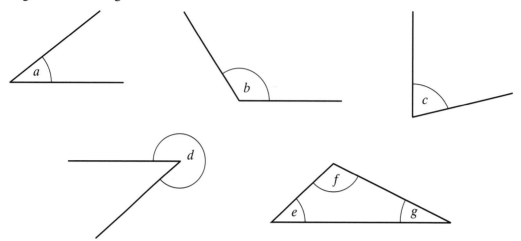

2 **Estimate** the size of each of the angles in question **1**.

3 **Measure** the size of each of the angles in question **1** with a protractor.
   Check that your answers agree with your answers to question **1**.

4 What is $e + f + g$?

Explore

◎ Draw a 4-sided figure (not a square or a rectangle)

◎ Measure the angles

◎ Add them up

◎ Repeat this process for 5-sided and 6-sided figures. What do you notice?

Investigate further

Learn 3 Angles and lines

Examples: Find the missing angles.

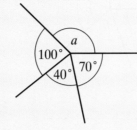

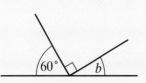

$$100° + 40° + 70° + a = 360°$$
$$210° + a = 360°$$
$$a = 150°$$

$$60° + 90° + b = 180°$$
$$150° + b = 180°$$
$$b = 30°$$

Angles at a point add up to 360°

Angles on a straight line add up to 180°

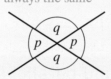

$$110° + c = 180°$$
$$c = 70°$$

Similarly
$$c + d = 180°$$
$$70° + d = 180°$$
$$d = 110°$$

Also
$$d + e = 180°$$
$$110° + e = 180°$$
$$e = 70°$$

Opposite angles are always the same

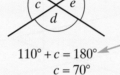

Apply 3

1 Calculate the size of each of the marked angles.
The diagrams are not drawn accurately.

a

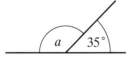

c

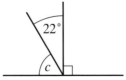

e

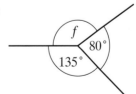

b

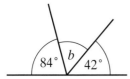

d

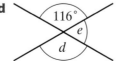

f

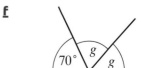

g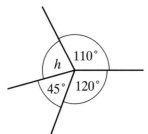

h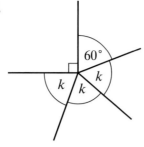

i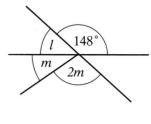

2 From the diagram below, find

   a two lines that are perpendicular to LM,

   b another pair of perpendicular lines,

   c two pairs of parallel lines.

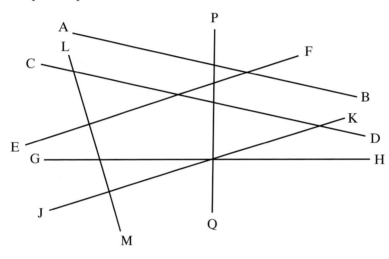

Learn 4 Angles and parallel lines

Examples: Find the missing angles.

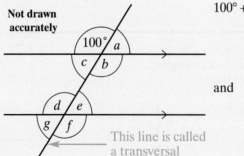

Not drawn accurately

This line is called a transversal

$100° + a = 180°$
$a = 80°$ Angles on a straight line add up to 180°

$b = 100°$ Opposite angles are equal

and

$c = a$
$c = 80°$

$d = 100°$ These angles are called corresponding angles   Corresponding angles are equal

Similarly

$e = a$ corresponding angles
$e = 80°$

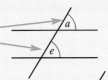

$f = b$ corresponding angles
$f = 100°$

$g = c$ corresponding angles
$g = 80°$

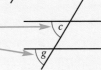

From the diagram $\left.\begin{array}{c} c = e \\ b = d \end{array}\right\}$ These are called alternate angles

Apply 4

Work out the size of each of these angles.
Write down a reason beside each answer.

Your reason will be one of these: **alternate angles**   **corresponding angles**

opposite angles   **angles on a straight line**

The diagrams are not drawn accurately.

1

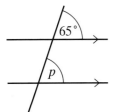

65°
p

3

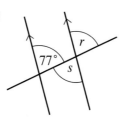

r
77°
s

5

130°
x
y

There are two steps
to find y, so give
two reasons.

2

38°
q

4

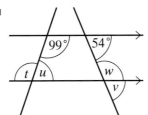

99°   54°
t u w
v

6

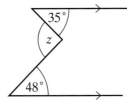

35°
z
48°

HINT   To find z put in another
parallel line.

Learn 5 Three-figure bearings

Examples: In each case, what is the bearing of A from B?

a

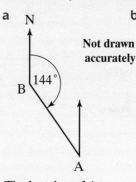

Not drawn accurately

The bearing of A from B is 144°.

b

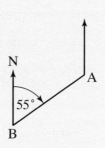

The bearing of A from B is 055°.

c

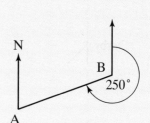

The bearing of A from B is 250°.

Bearings are measured from North in a clockwise direction

Bearings are always written as 3 figures, so 55° is written 055°

Apply 5

1 For each diagram, write down the 3-figure bearing of D from E.

a

N

E 115°

D

d

N

D

74°

E

b

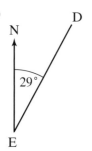

D

N

29°

E

Not drawn accurately

e

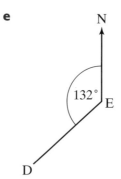

N

132° E

D

c

N

D —— E

f

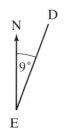

N D

9°

E

2 Get Real!

The diagram shows the three towns of Ashby, Derby and Loughborough.
Derby is due north of Ashby.
Loughborough is on a bearing of 080° from Ashby.
The bearing of Loughborough from Derby is 137°.

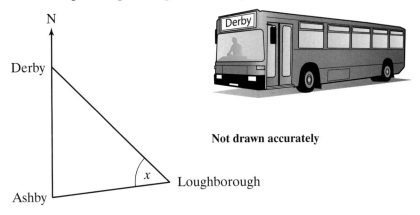

Not drawn accurately

Copy the diagram, adding the bearings and working out the size of angle x.

3 Get Real!

The diagram shows the three towns of Nelson, Haworth and Todmorden.
Haworth is due east of Nelson.
The bearing of Todmorden from Nelson is 150°.
The bearing of Todmorden from Haworth is 220°.

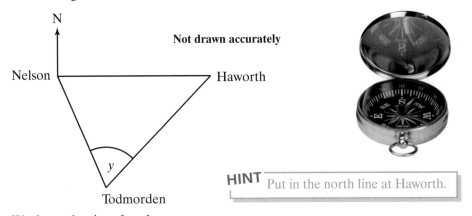

Not drawn accurately

HINT Put in the north line at Haworth.

Work out the size of angle y.

Angles

The following exercise tests your understanding of this chapter,
with the questions appearing in order of increasing difficulty.

1 Calculate the number of degrees in:

    a $\frac{2}{3}$ of a revolution   **c** 0.3 of a revolution

    b $\frac{1}{8}$ of a revolution   **d** $1\frac{1}{2}$ revolutions

2 The second hand on a clock revolves at 1 revolution per minute.
How many degrees per second is this?

3 What is the angle between the hour hand and the minute hand of a clock

    a at two o'clock   **b** at half past five?

4 Calculate the size of each of the marked angles.

a

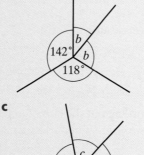

d

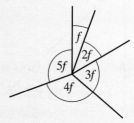

b

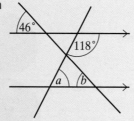

e

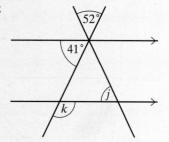

c

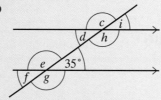

                                           Not drawn accurately

5 Calculate the size of each of the marked angles.

a

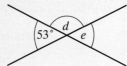

c

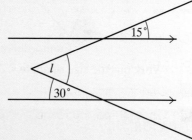

b

d

    Not drawn accurately

> **HINT** Draw another parallel line.

6 The cruise liner 'Oriana' is sailing on a bearing of 060°.
    To avoid a storm ahead it changes course to a bearing of 125°.
    Through what angle has it turned?

7 Two aircraft take off from Birmingham International Airport, one on a
    bearing of 152° and the other on a bearing of 308°.
    What is the angle between the two planes' flight paths?

3 Integers

OBJECTIVES

G → **Examiners would normally expect students who get a G grade to be able to:**

Understand positive and negative integers

Find the factors of a number

F → **Examiners would normally expect students who get an F grade also to be able to:**

Add and subtract negative integers

E → **Examiners would normally expect students who get an E grade also to be able to:**

Multiply and divide negative integers

C → **Examiners would normally expect students who get a C grade also to be able to:**

Recognise prime numbers

Find the reciprocal of a number

Find the least common multiple (LCM) of two simple numbers

Find the highest common factor (HCF) of two simple numbers

Write a number as a product of prime factors

What you should already know ...

- Understand the four rules of number
- Understand place value
- Understand the inequality signs <, >, ≤ and ≥

- Know the meaning of 'sum' and 'product'
- Change a decimal into a fraction
- Change a mixed number into a top-heavy fraction

VOCABULARY

Counting number or **natural number** – a positive whole number, for example, 1, 2, 3, ...

Positive number – a number greater than 0; it can be written with or without a positive sign, for example, 1, +4, 8, 9, +10, ...

Negative number – a number less than 0; it is written with a negative sign, for example, −1, −3, −7, −11, ...

Integer – any positive or negative whole number or zero, for example, −2, −1, 0, 1, 2, ...

Directed number – a number with a positive or negative sign attached to it; it is often seen as a temperature, for example, −1, +1, +5, −3°C, +2°C, ...

Less than (<) – the number on the left-hand side of the sign is smaller than that on the right-hand side

Greater than (>) – the number on the left-hand side of the sign is larger than that on the right-hand side

Sum – to find the sum of two numbers, you add them together

Product – the result of multiplying together two (or more) numbers, variables, terms or expressions

Factor – a natural number which divides exactly into another number (no remainder); for example, the factors of 12 are 1, 2, 3, 4, 6, 12

Multiple – the multiples of a number are the products of the multiplication tables, for example, the multiples of 3 are 3, 6, 9, 12, 15, ...

Least common multiple (LCM) – the least multiple which is common to two or more numbers, for example,

the multiples of 3 are 3, 6, 9, 12, 15, 18, 24, 27, 30, 33, 36, ...
the multiples of 4 are 4, 8, 12, 16, 20, 24, 28, 32, 36, ...
the common multiples are 12, 24, 36, ...
the least common multiple is 12

Common factor – factors that are in common for two or more numbers, for example,

the factors of 6 are 1, 2, 3, 6
the factors of 9 are 1, 3, 9
the common factors are 1 and 3

Highest common factor (HCF) – the highest factor that two or more numbers have in common, for example,

the factors of 16 are 1, 2, 4, 8, 16
the factors of 24 are 1, 2, 3, 4, 6, 8, 12, 24
the common factors are 1, 2, 4, 8
the highest common factor is 8

Prime number – a natural number with exactly two factors, for example, 2 (factors are 1 and 2), 3 (factors are 1 and 3), 5 (factors are 1 and 5), 7, 11, 13, 17, 23, ..., 59, ...

Index notation – when a product such as $2 \times 2 \times 2 \times 2$ is written as 2^4, the number 4 is the index (plural **indices**)

Reciprocal – any number multiplied by its reciprocal equals one; one divided by a number will give its reciprocal, for example,

the reciprocal of 3 is $\frac{1}{3}$ because $3 \times \frac{1}{3} = 1$

Learn 1 Positive and negative integers

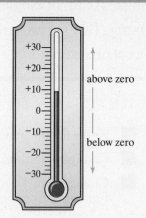

Directed numbers are used to show temperatures above and below zero

Example: Put these numbers in ascending order.

+3, −8, −12, +20, −24, −2

In ascending order the numbers are

−24, −12, −8, −2, +3, +20

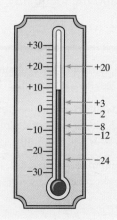

Apply 1

1 Put these numbers in ascending order:

   a $+3, 0, -7, +6, +1, +10, -5, -1, +9$

   b $-4, -9, +8, 9, 12, +6, -7, -3, 0$

   c $67, -9, +78, -98, 876, -987, 634$

2 Write down a number that is less than each of these.

   a -4 **b** -3 **c** -6 **d** -7

3 A thermometer shows 5 degrees below freezing. Write this as a directed number.

4 Get Real!

Clare is £162 overdrawn at the bank. Write this as a directed number.

5 Get Real!

In Plymouth the temperature was $-1°C$. In Manchester on the same night, the temperature was $-4°C$. Which city was warmer and by how much?

6 Put the correct sign, $<$, $>$ or $=$, between these numbers.

   a $-4 \square -9$ **b** $-3 \square -2$ **c** $-6 \square 4$ **d** $5 \square -5$ **e** $6 \square +6$

7 Alison is going sailing.

She starts from her home which is 300 m above sea level. She walks to the marina where she gets her boat.

After setting sail she decides to drop anchor and swim. She dives in and goes 5 m under water. She sees a turtle swimming.

What is the height difference between her house and the turtle?

8 Make up your own story to include positive and negative numbers. Draw a diagram to show the information.

9 Find the largest four-figure number that can be made using each of the numbers once.

   a 3 4 8 2   **b** 1 5 9 6   **c** 2 6 7 8

10 Find the smallest four-figure number that can be made using each of the numbers once.

   a 3 4 8 2   **b** 1 5 9 6   **c** 2 6 7 8

11 Find all the four-figure numbers that can be made using each of the digits once. Arrange them in ascending order.

   a 4 5 2 1   **b** 3 8 6 4   **c** 1 7 9 2

12 Get Real!

You want as much money as possible in your bank account. Put these amounts in order of preference.

a £140, £20 (overdrawn), £150 (credit), £200 (overdrawn), £30 (credit)

b £200, −£150, £30 (credit), −£125, £125 (credit)

c £145 (overdrawn), −£135, £245, £30 (credit), £57 (overdrawn)

Learn 2 Adding and subtracting negative integers

Examples:

a What is the value of + 1 + + 4

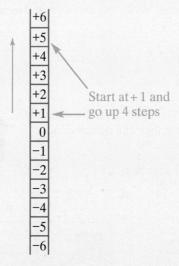

c What is the value of 3 + − 2?

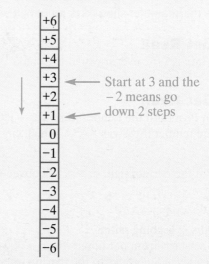

b What is the value of − 5 + + 6

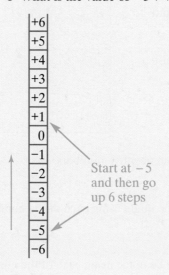

d What is the value of − 2 − − 4?

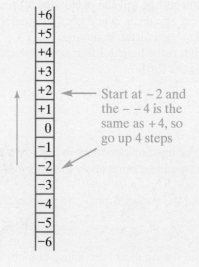

Apply 2

1 Copy this addition table. Starting at the top right-hand corner,
   calculate $A + B$ and fill in the boxes.
   Use the patterns you find to complete the table.

					B				
$A+B$	−4	−3	−2	−1	0	+1	+2	+3	+4
+4									
+3									
+2									
+1									
0									
−1									
−2									
−3									
−4									

(A labels the rows)

2 Find the missing number in each of the following.

   a $+3 + \ldots = +5$ c $-1 + \ldots = -3$ e $4 + \ldots = -9$

   b $+3 + \ldots = 0$ d $-2 + \ldots = -1$ f $\ldots + 1 = -3$

3 What must be added to:

   a −4 to make 6?   b 2 to make −2?   c 13 to make 3?

4 Andrew says that $-3 + 5 = -8$. Is he correct? Explain your answer.

5 Copy this subtraction table. Starting at the top right-hand corner,
   calculate $A - B$ and fill in the boxes.
   Use the patterns you find to complete the table.

					B				
$A-B$	−4	−3	−2	−1	0	+1	+2	+3	+4
+4									
+3									
+2									
+1									
0									
−1									
−2									
−3									
−4									

(A labels the rows)

6 Find the missing number in each of the following.

   a $+3 - \ldots = +1$ c $-3 - \ldots = -2$ e $-3 - \ldots = 0$

   b $+4 - \ldots = +5$ d $-1 - \ldots = 2$ f $\ldots - -2 = 3$

7 What must be subtracted from:

a 14 to make 7   **b** -2 to make -9 **c** -6 to make -3?

8 Get Real!

a If the temperature is $+3°C$ and it falls by $9°C$, what is the new temperature?

b If the temperature is $-6°C$, by how many degrees must it rise to become $6°C$?

c After rising $9°C$ the temperature is $+3°C$. What was it originally?

9 a Fill in the boxes so that the answer to each question is 12.

   i $15 - \boxed{} = 12$ **iii** $-2 + \boxed{} = 12$

   ii $\boxed{} + -6 = 12$ **iv** $\boxed{} - +8 = 12$

b Make up five more addition and subtraction questions that have 12 as an answer. You must use negative and positive numbers in each question.

Learn 3   Multiplying and dividing negative integers

Examples: Work out the answers to the following questions.

   a $+2 \times +3$ **c** $-2 \times +3$ **e** $-12 \div +4$

   b $+2 \times -3$ **d** -2×-3 **f** $-6 \div -3$

   a $+2 \times +3$ ⟵ Remember that $+2 \times +3$ **Signs the same = positive**
       $= +6$ is the same as 2×3

   b $+2 \times -3$ ⟵ This is the same as 2×-3 **Signs different = negative**
       $= -6$

   c $-2 \times +3$ ⟵ This is the same as -2×3 **Signs different = negative**
       $= -6$

   d -2×-3 ⟵ Not -6 **Signs the same = positive**
       $= +6$

   e $-12 \div +4$ ⟵ The same rules apply as for multiplication
       $= -3$ **Signs different = negative**

   f $-6 \div -3$ ⟵ The same rules apply as for multiplication
       $= 2$ **Signs the same = positive**

Apply 3

1 Copy the multiplication table. Starting at the top right-hand corner,
calculate $A \times B$ and fill in the boxes.
Use the patterns you find to complete the table.

$A \times B$	-4	-3	-2	-1	0	$+1$	$+2$	$+3$	$+4$
$+4$									
$+3$									
$+2$									
$+1$									
0									
-1									
-2									
-3									
-4									

(B labels the columns, A labels the rows.)

2 Find the missing number in each of the following:

a $+3 \times \ldots = +15$

b $+3 \times \ldots = -30$

c $-1 \times \ldots = -3$

d $-2 \times \ldots = -12$

e $4 \times \ldots = -36$

f $\ldots \times -10 = 30$

g $-16 \div \ldots = -4$

h $\ldots \div -4 = -8$

i $\ldots \div -2 = 8$

3 Find the values of these:

a $\dfrac{-2 \times +9}{-3}$

b $\dfrac{16 \times -3}{+4}$

c $\dfrac{-6 \times -7}{-2}$

4 The answer is 24. Can you make up ten questions, involving multiplication
and division, that give that answer? You must use negative numbers.

5 Grace has been asked to write down the first five terms of a sequence which starts with 1.
The rule for finding the next number is 'multiply the last number by -2'. She writes:

$\quad 1 \quad -2 \quad -4 \quad -8 \quad -16$

Is she correct? Explain your answer.

6 Find

a two numbers whose sum is -8 and whose product is 15

b two numbers whose sum is -1 and whose product is -20.

Explore

◎ What happens if you keep multiplying negative numbers?

◎ For example, multiply two negative numbers, the answer is positive

◎ Multiply three negative numbers, the answer is ...

Investigate further

Learn 4   Factors and multiples

Examples: **a** What are the factors of 28?

$$1 \quad 2 \quad 4 \quad 7 \quad 14 \quad 28$$

In most cases factors are in pairs
$1 \times 28 = 28$
$2 \times 14 = 28$
$4 \times 7 = 28$

b What are the first five multiples of 4?

$$4 \quad 8 \quad 12 \quad 16 \quad 20$$

All of the numbers are in the four times table

c What is the least common multiple (LCM) of 6 and 8?

$$6 \quad 12 \quad 18 \quad 24 \quad 30$$ ——These numbers are all multiples of 6

$$8 \quad 16 \quad 24 \quad 32 \quad 40$$ ——These numbers are all multiples of 8

The LCM of 6 and 8 is 24.   24 is the smallest number that is common to both lists

d What is the highest common factor (HCF) of 16 and 24?

$$1 \quad 2 \quad 4 \quad 8 \quad 16$$ ——These numbers are all factors of 16

$$1 \quad 2 \quad 3 \quad 4 \quad 6 \quad 8 \quad 12 \quad 24$$ ——These numbers are all factors of 24

$$1 \quad 2 \quad 4 \quad 8$$ ——1, 2, 4, 8 are common factors of 16 and 24

The HCF of 16 and 24 is 8.   8 is the highest number that is common to both lists

Apply 4

1 From this set of numbers

$$2 \quad 3 \quad 4 \quad 5 \quad 6 \quad 7 \quad 8 \quad 9$$

write down the numbers that are factors of:

a 6   **c** 4   **e** 25

b 9   **d** 16   **f** 24

2 Write down all the factors of:

a 15   **c** 48   **e** 40   **g** 32   **i** 84

b 64   **d** 10   **f** 36   **h** 72

3 Find the common factors of:

a 6 and 15   **c** 4 and 64   **e** 25 and 40

b 9 and 48   **d** 10 and 16   **f** 24 and 36

4 Find the HCF of the following sets of numbers.

a 6 and 15   **d** 24 and 36   **g** 84 and 70

b 12 and 15   **e** 27 and 36

c 32 and 48   **f** 56 and 152

5 Write down all the factors of 20 and 24. Hence find the common factors and write down the HCF of 20 and 24.

6 The HCF of two numbers is 5. Give five possible pairs of numbers.

7 Write down the first five multiples of:

a 2	**c** 7	**e** 9	**g** 11	**i** 13
b 5	**d** 6	**f** 12	**h** 8	

8 Find the LCM of these sets of numbers.

a 6 and 15	**d** 3 and 8	**g** 3, 5 and 6
b 12 and 6	**e** 4 and 6	**h** 6, 8 and 32
c 5 and 7	**f** 4, 10 and 12	

9 Get Real!

A lighthouse flashes every 56 seconds. Another lighthouse flashes every 40 seconds. At 9 p.m. they both flash at the same time. What time will it be when they next both flash at the same time?

10 Get Real!

Alison is making her own birthday cards. She needs to cut up lengths of ribbon. Find the smallest length of ribbon that can be cut into an exact number of either 5 cm or 8 cm or 12 cm lengths.

11 Get Real!

One political party holds its annual conference at Eastbourne every four years. Another holds its annual conference there every six years. They both held their conference in Eastbourne in 2006. When will they next be there in the same year?

12 Get Real!

A rectangular floor measures 450 cm by 350 cm. What is the largest square tile that can be used to cover the floor without any cutting?

13 Get Real!

Rectangular tiles measure 15 cm by 9 cm. What is the length of the side of the smallest square area that can be covered with these tiles?

Explore

◎ Write down all the numbers between 1 and 30

◎ Work out the number of factors for each number

◎ Can you work out a rule for numbers that have

a two factors only

b an odd number of factors?

Investigate further

Learn 5   Prime numbers and prime factor decomposition

Example:

Find the prime factors of 40.

Start by finding the smallest prime number that divides into 40.
Continue dividing by successive prime numbers until the answer becomes 1.

2	40
2	20
2	10
5	5
	1

Prime numbers are numbers
with exactly two factors, for
example, 2, 3, 5, 7, 11, 13, ...

The prime factors of 40 are 2, 2, 2, 5.
40 written as a product of prime factors is $2 \times 2 \times 2 \times 5$.
This can be written as $2^3 \times 5$.

This is called 'index notation'.
The index tells you how many
times the factor 2 occurs

Apply 5

1   Write each of the following numbers as a product of prime factors.

   a 20   **c** 36   **e** 90   **g** 63   **i** 84

   b 18   **d** 66   **f** 100   **h** 48   **j** 96

2   Express each number as a product of its prime factors.

   a 24   **b** 72   **c** 45

3   Express each number as a product of its prime factors.
    Write your answers using index notation.

   a 220   **c** 136   **e** 720   **g** 390   **i** 624

   b 144   **d** 300   **f** 480   **h** 450   **j** 216

4   Clare says that one must be a prime number. Is she correct? Explain your answer.

5   What number is this?

   a It is less than 100, it is 1 less than a multiple of 7, it is a prime number and
      its digits add up to 5.

   b It is less than 100, it is a multiple of 11 and it is 2 more than a square number.
      If it is divided by 9 there is a remainder of 3.

6   Write 2420 as a product of its prime factors. Write your answer using index notation.

7   Write 9240 as a product of its prime factors. Write your answer using index notation.

8   Write 8820 as a product of its prime factors. Write your answer using index notation.

Explore

You will need a 100 square

◎ Cross out the number 1

◎ Put a circle round the number 2 and then cross out all of the other multiples of 2

◎ Put a circle round the next number after 2 which has not been crossed out

◎ Cross out all of the other multiples of that number

◎ Put a circle round the next number not crossed out and cross out every multiple of that number

◎ Continue until you run out of numbers in the 100 square

What do you notice about the numbers that are left?

Investigate further

Learn 6 Reciprocals

Examples: Find the reciprocal of: **a** 5   **b** $\frac{1}{4}$ **c** 0.3   **d** $2\frac{1}{2}$

a The reciprocal of 5 is $\frac{1}{5}$ ← — You can write 5 as $\frac{5}{1}$
The reciprocal of $\frac{5}{1}$ is $\frac{1}{5}$

b The reciprocal of $\frac{1}{4}$ is $\frac{4}{1} = 4$ ← — It is better to write $\frac{4}{1}$ as 4

c The reciprocal of 0.3 is the same as ← — Write 0.3 as a fraction
the reciprocal of $\frac{3}{10}$

The reciprocal of $\frac{3}{10}$ is $\frac{10}{3} = 3\frac{1}{3}$ ← — It is better to write $\frac{10}{3}$ as $3\frac{1}{3}$

d The reciprocal of $2\frac{1}{2}$ is the same as ← — Write $2\frac{1}{2}$ as $\frac{5}{2}$
the reciprocal of $\frac{5}{2}$

The reciprocal of $\frac{5}{2}$ is $\frac{2}{5}$

Apply 6

 1 Write down the reciprocal of

    a 4   **b** 6   **c** 8   **d** 10   **e** 7   **f** 0.25

 2 Find the reciprocal of

    a $\frac{1}{2}$ **b** $\frac{1}{5}$ **c** $\frac{1}{7}$ **d** $\frac{1}{8}$ **e** $\frac{1}{12}$ **f** 0.8

3 Find the reciprocal of

    a $\frac{2}{7}$ **b** $\frac{3}{5}$ **c** $\frac{2}{3}$ **d** $\frac{5}{6}$ **e** 0.125   **f** $0.\dot{3}$

4 Find the reciprocal of

  a $2\frac{1}{4}$ **b** $3\frac{1}{2}$ **c** $1\frac{3}{4}$ **d** 1.25   **e** 3.6   **f** $1.\dot{6}$

5 Find the reciprocals of the numbers 2 to 12, as decimals. If they are not exact, write them as recurring decimals. Which of the numbers have reciprocals that

  a are exact decimals

  b have one recurring figure

  c have two recurring figures?

Explore

◉ Write down the reciprocals of $2, 1, \frac{1}{2}, \frac{1}{4}, \frac{1}{8}, \dots$

◉ Continue the pattern

◉ What do you notice?

> Investigate further

Integers

ASSESS

The following exercise tests your understanding of this chapter, with the questions appearing in order of increasing difficulty.

1 In a magic square each number is different. The sum of each row, each column and each diagonal is the same. Fill in the missing numbers in the magic square.

−1	−2	
	0	−4
	2	

2 In a magic square each number is different. The sum of each row, each column and each diagonal is the same.

	−9	−2	1	12
9	5	−6	7	
−10	4	2	0	14
	−3	10	−1	−7
−8		6		−4

  a Fill in the missing numbers.

  b What is the sum of any row, column or diagonal?

  This magic square has a smaller magic square inside it.

  c Find this magic square.

  d What is the sum of each of its diagonals, rows and columns?

3 Copy and complete the following table.

Temperature	Change	New temperature
4	+5	
4	−7	
−2	+6	
−1	−4	
13		22
24		1
1		−3
−2		−4
	+4	7
	+13	9
	−11	−2
	−7	2

4 Julius Sneezar was born in the year 25 BC and his wife, Bigga, in 21 BC.
Julius died in AD 33 and Bigga in AD 41.
Assume that they had each had their birthdays in the year they died.

    a How old was Julius when he died?

    b How old was Bigga when she died?

5 Find the values of each of these.

    a 2×-5 **d** -4×0 **g** $-45 \div -15$

    b -6×7 **e** $16 \div -4$ **h** $0 \div -2$

    c -8×-3 **f** $-24 \div 2$

6 Write down the first six terms of the sequence that starts with 100
and where each term is the previous term divided by -2.

7 At a party it was discovered that Siobhan, Gareth, Nathan and Ulrika had
birthdays on the 6th, 15th, 27th and 30th of the month. Sven joined the
group and it was discovered that his birthday was a factor of everyone else's.
If Sven was not born on the 1st, on what day of the month was Sven born?

8 Find the prime factors of: **a** 420   **b** 13 475

9 Find the reciprocals of:

    a 5   **c** $\frac{5}{8}$ **e** $0.\dot{2}$

    b -8 **d** -0.2

10 a What is the only number that is the same as its reciprocal?

    b What is the only number that has no reciprocal?
Explain your answer.

Try some real past exam questions to test your knowledge:

11 Tom, Sam and Matt are counting drum beats.

Tom hits a snare drum every 2 beats.
Sam hits a kettle drum every 5 beats.
Matt hits a bass drum every 8 beats.

Tom, Sam and Matt start by hitting their drums at the same time.
How many beats is it before Tom, Sam and Matt **next** hit their drums at the
same time?

Spec A, Higher Paper 1, June 04

12 a Express 144 as the product of its prime factors.
Write your answer in index form.

b Find the Highest Common Factor (HCF) of 60 and 144.

Spec B, Mod 3 Intermediate, June 03

4 Rounding

G   **Examiners would normally expect students who get a G grade to be able to:**

Round to the nearest integer

Write an integer correct to the nearest 10 or the nearest 100

Estimate answers to problems involving decimals

F   **Examiners would normally expect students who get an F grade also to be able to:**

Estimate square roots

Round numbers to given powers of 10 and to a given number of decimal places

E   **Examiners would normally expect students who get an E grade also to be able to:**

Round a number to one significant figure

D   **Examiners would normally expect students who get a D grade also to be able to:**

Estimate answers to calculations such as $\dfrac{22.6 \times 18.7}{5.2}$

C   **Examiners would normally expect students who get a C grade also to be able to:**

Estimate answers to calculations such as $\dfrac{22.6 \times 18.7}{0.52}$

Find minimum and maximum values

What you should already know ...

■ Arrange whole numbers and decimal numbers in order of size

■ Work with number lines

Round – give an approximate value of a number; numbers can be rounded to the nearest 1000, nearest 100, nearest 10, nearest integer, significant figures, decimal places, ... etc.

Significant figures – the digits in a number; the closer a digit is to the beginning of a number then the more important or significant it is; for example, in the number 23.657, 2 is the most significant digit and is worth 20, 7 is the least significant digit and is worth $\frac{7}{1000}$; the number 23.657 has 5 significant digits

Decimal places – the digits to the right of a decimal point in a number, for example, in the number 23.657, the number 6 is the first decimal place (worth $\frac{6}{10}$), the number 5 is the second decimal place (worth $\frac{5}{100}$) and 7 is the third decimal place (worth $\frac{7}{100}$); the number 23.657 has 3 decimal places

Estimate – find an approximate value of a calculation; this is usually found by rounding all of the numbers to one significant figure, for example, $\frac{20.4 \times 4.3}{5.2}$ is approximately $\frac{20 \times 4}{5}$ where each number is rounded to 1 s.f., the answer can be worked out in your head to give 16

Upper bound – this is the maximum possible value of a measurement, for example, if a length is measured as 37 cm correct to the nearest centimetre, the upper bound of the length is 37.5 cm

Lower bound – this is the minimum possible value of a measurement, for example, if a length is measured as 37 cm correct to the nearest centimetre, the lower bound of the length is 36.5 cm

Learn 1　Rounding numbers and quantities

Example:

Round 63 and 48 to the nearest 10.

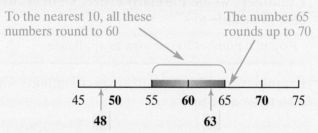

To the nearest 10, all these numbers round to 60

The number 65 rounds up to 70

All numbers between 55 and 65 round to 60 to the nearest 10, because they are nearer to 60 than they are to 50 or 70.

So 63 rounded to the nearest 10 is 60.

48 is nearer to 50 than 40, so 48 rounded to the nearest 10 is 50.

Apply 1

1

Round these to the nearest whole number.
(The number line may help to get you started.)

  a 7.6   **b** 5.2   **c** 67.8   **d** 0.8   **e** 0.2   **f** 89.5

2 Round these numbers   **a** to the nearest 10   **b** to the nearest 100.

  i 166   **ii** 234   **iii** 2022   **iv** 1598   **v** 16.1   **vi** 245

3 Round the numbers in question **2** to the nearest 5.

4 a Hannah says that 3284 rounded to the nearest 100 is 33.
What has she done wrong?

b Sanjay says that 7.6 rounded to the nearest whole number is 8.0
What mistake has he made?

5 Here is some incorrect rounding. Write a corrected statement in each case.

a 416 291 rounds to 4163

b 2997 to the nearest 10 is 2990

c 54.8 to the nearest whole number is 55.0

6 Draw a number line marked with numbers from 0 to 5 and colour the part of the line

a containing all the numbers that round to 3 when rounded to the nearest whole number

b containing all the numbers that round to 1.4 when rounded to the nearest tenth.

7 Write down four different 4-digit numbers that

a become 9300 when rounded to the nearest 100

b become 9300 when rounded to the nearest 50.

c Which of your numbers could be the answers to both parts **a** and **b**?

8 Get Real!
What is the length of this pencil **a** to the nearest centimetre **b** to the nearest half centimetre
c to the nearest 5 centimetres **d** to the nearest 10 centimetres?

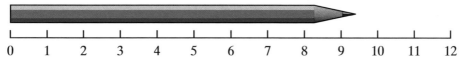

```
|   |   |   |   |   |   |   |   |   |   |   |   |   |
0   1   2   3   4   5   6   7   8   9   10  11  12
```

9 Aled's height is 148 cm. What is this to the nearest 10 cm? To the nearest half metre?

10 Round these amounts of money **a** to the nearest pound **b** to the nearest 50 pence.

i £54.64   **ii** £1.73   **iii** £235.24   **iv** 84 pence   **v** £0.34

11 An amount of money correct to the nearest pound is £52.
What is the largest amount it could be? What is the smallest?

12 The number of people at a concert is 7000 to the nearest 1000 people.
What is the smallest possible number of people at the concert?

13 The number of spectators at a football match was 34 485.
A local newspaper reported this as 35 000.
Was this correct? Give a reason for your answer.

Explore

◎ Mathematicians and statisticians don't always round numbers like 15.5 and 4.5 up to the next whole number above

◎ Instead, they round up or down to make the number **even** – so 15.5 rounds to 16 and 4.5 rounds to 4

◎ Why might this be a good idea?

Investigate further

Learn 2 Rounding to significant figures and decimal places

Examples:

a Round the following numbers to:

   i one significant figure   **ii** two significant figures.

   3456 345.6 34.56 3.456 0.3456 0.03456 0.003456

Number	One significant figure	Two significant figures
3456	3000	3500
345.5	300	350
34.56	30	35
3.456	3	3.5
0.3456	0.3	0.35
0.03456	0.03	0.035
0.003456	0.003	0.0035

Zeros are not significant figures. They are used to pad out the number to make it the correct size

3.456 is closer to 3.5 than 3.4 when written to one significant figure

The number 3 is the most significant figure when rounding to one significant figure

The numbers 3 and 5 are the most significant figures when rounding to two significant figures

b Round the following numbers to:

   i one decimal place   **ii** two decimal places.

   34.56 3.456 0.3456 0.03456 0.003456

Number	One decimal place	Two decimal places
34.56	34.6	34.56
3.456	3.5	3.46
0.3456	0.3	0.35
0.03456	0.0	0.03
0.003456	0.0	0.00

3.456 is closer to 3.5 than 3.4 when written to one decimal place

0.3456 is closer to 0.35 than 0.34 when written to two decimal places

In this example the answers contain only one decimal place

In this example the answers contain only two decimal places

Apply 2

1 Round these   **a** to one significant figure   **b** to two significant figures.

   i 166   **ii** 234   **iii** 2022   **iv** 1598   **v** 16.1   **vi** 245

2 Round these   **a** to one significant figure   **b** to two significant figures.

   i 3.78   **ii** 0.378   **iii** 0.0526   **iv** 0.00526   **v** 0.000526   **vi** 0.0000526

3 Round the numbers in question **2** to two decimal places.

4 Get Real!
Round these amounts of money to two decimal places
(that is, to the nearest penny).

   a £19.257   **b** £25.387   **c** £235.24   **d** £25.397   **e** £0.5621

5 Round these measurements to one decimal place (that is, to the nearest millimetre).

    a 24.67 cm   **b** 4.93 cm   **c** 73.45 cm   **<u>d</u>** 0.566 cm   **<u>e</u>** 0.5454 cm

<u>6</u> Round these weights to three decimal places (that is, to the nearest gram).

    a 1.5396 kg   **b** 41.733 kg   **c** 5.9863 kg   **d** 0.045365 kg   **e** 0.0008454 kg

Explore

◎ Find a number that is the same when rounded to the nearest whole number as to one significant figure

◎ Find a number that is the same when rounded to the nearest 10 as to one significant figure

◎ Find a number that is the same when rounded to one decimal place as to one significant figure

◎ Find a number that is the same when rounded to two decimal places as to one significant figure

Investigate further

Learn 3   Estimating

Example: Estimate the answer to 3.86×2.14

Round all the numbers to one significant figure, then work out the approximate answer in your head.

3.86 is 4 to one significant figure   ⟶ 3.86×2.14 ⟵   2.14 is 2 to one significant figure

$$3.86 \times 2.14 \approx 4 \times 2 = 8$$

This curly equals sign means 'is approximately equal to'

It is easy to make mistakes when using decimals, so it is a good idea to estimate to find the approximate size of the answer so that you can see if you are right.

Apply 3

1 For each question, decide which is the best estimate.

		Estimate A	Estimate B	Estimate C
a	2.89×9.4	2.7	18	27
b	1.2×29.4	3	30	300
c	$9.17 \div 3.2$	3	4	18
d	48.5×9.8	5	50	500
e	$4.2 \div 1.9$	1	2	3
f	22.4×6.1	12	120	180
g	$7.8 \div 1.2$	8	78	80
h	$2.1 \times 3.1 \div 4.2$	1	1.5	2
i	$20.9 \div 6.9 \times 4.1$	10	11	12

2 Estimate the answers to these calculations by rounding to one significant figure. You may wish to use a calculator to check your answers.

a $2.9 + 3.2$

f $4.3 - 3.7$

k $\dfrac{102.4 + 8.7}{0.22}$

b $7.9 \div 2.2$

g 5.3×8.2

l $\dfrac{102.4 \times 8.7}{0.22}$

c $67.8 + 22.1$

h $\dfrac{20.4 \times 7.7}{0.52}$

m $21.3(7.56 + 3.89)$

d $\dfrac{20.4 \times 7.7}{5.2}$

i $\dfrac{75.5 \times 2.7}{0.12}$

n $21.3(7.56 - 3.89)$

e 0.2×5.4

j $\dfrac{28.5 + 53}{64.1 - 53.7}$

HINT Be careful when dividing by numbers less than one.

3 For each pair of calculations, estimate the answers to help decide which you think will have the bigger answer. Use a calculator to check your answers.

a 5.2×1.8 or 3.1×2.95

c $9.723 + 4.28$ or $39.4 \div 2.04$

b $28.4 \div 5.9$ or 2.03×3.78

d 39.5×21.3 or 81.3×7.8

4 The square root of 10 lies between the square root of 9 and the square root of 16. The square root of 9 is 3 and the square root of 16 is 4, so the answer lies between 3 and 4.

Use this method to estimate the square root of:

a 12   **b** 20   **c** 50   **d** 3   **e** 1000

5 Estimate 5.92×3.82 by rounding to the nearest whole number. Explain why the answer is an over-estimate of the exact answer.

6 Sam says, 'I estimated the answer to $16.7 - 8.6$ as 8 by rounding up both numbers, so the answer is an over-estimate.' Show that Sam is not correct.

7 Ali says, '35 divided by 5 is 7, so 35 divided by 0.5 is 0.7'. Is Ali right? Give a reason for your answer.

8 Give one example that shows that dividing can make something smaller and another to show that dividing can make something bigger.

9 Hannah says:

- When I rounded the numbers in a calculation I got $\dfrac{110}{0.2}$
- Then I multiplied the top and the bottom both by 10 to give $\dfrac{1100}{2}$
- So the answer is 550.

Is Hannah correct? Give a reason for your answer.

10 a Find five numbers whose square roots are between 7 and 8.

b Find two consecutive whole numbers to complete this statement: 'The square root of 60 is between ... and ...'

11 Get Real!
Estimate the total cost of three books costing £3.99, £5.25 and £10.80

12 Get Real!

Estimate the length of fencing needed for this field.

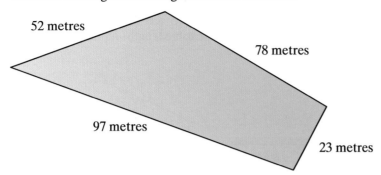

52 metres

78 metres

97 metres

23 metres

13 Get Real!

Anne's car goes 6.2 miles on every litre of petrol. Estimate how far she can drive if her fuel tank has 24.5 litres in it.

14 Get Real!

A group of 18 people wins £389 540 on the lottery. Estimate how much each person will get when the money is shared out equally.

15 Get Real!

It's Harry's birthday! He asks his mum for a cake like a football pitch. She makes a cake that is 29 cm wide and 38 cm long.

a She wants to put a ribbon round the cake. She can buy ribbon in various lengths: 1 m, 1.5 m, 2 m, 2.5 m or 3 m.
Estimate the perimeter of the cake and say which length ribbon she should buy.

b She can buy ready-made green icing for the top of the cake. The icing comes in packs to cover 1000 cm^2.
Estimate the area of the top of the cake and decide whether one pack will be enough.

Explore

 You know that $20 \times 30 = 600$
So the answers to all these calculations will be close to 600, as the numbers are close to 20×30:

a 19.4×28.7 **d** 21.2×30.4 **g** 23.4×33.4
b 23.4×30.2 **e** 29.8×20.4 **h** 22.3×29.8
c 18.8×29.6 **f** 31.4×18.7

 Can you decide which answers will be less than 600, and which will be more than 600?

 Check your predictions with a calculator

 When can you be sure an estimate is lower than the actual answer?

 When can you be sure an estimate is higher than the actual answer?

Investigate further

Learn 4   Finding minimum and maximum values

Example:

The length of a table is measured as 60 cm, correct to the nearest centimetre. What are the minimum and maximum possible lengths of the table?

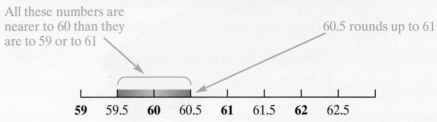

All these numbers are nearer to 60 than they are to 59 or to 61

60.5 rounds up to 61

59   59.5   **60**   60.5   **61**   61.5   **62**   62.5

Any length given to the nearest centimetre could be up to half a centimetre smaller or larger than the given value

A length given to the nearest centimetre as 60 cm could be anything between 59.5 cm and 60.5 cm. The minimum and maximum possible lengths are 59.5 cm and 60.5 cm or $59.5 \leqslant x < 60.5$

The length cannot actually be 60.5 cm, as this measurement rounds up to 61 cm – but it can be as close to 60.5 cm as you like, so 60.5 cm is the top limit (or **upper bound**) of the length

Apply 4

1 Each of these quantities is rounded to the nearest whole number of units. Write down the minimum and maximum possible size of each quantity.

a 54 cm

d 17 mℓ

b 5 kg

e £45

c 26 m

f 175 g

2 Jane says,

'If a length is 78 cm to the nearest centimetre, then the maximum possible length is 78.49 cm.'

Is Jane right? Explain your answer.

3 The volume of water in a tank is given as 1500 litres.

a Decide if the volume has been rounded to the nearest litre, nearest 10 litres or nearest 100 litres if the minimum possible volume is:

i 1450 ℓ **ii** 1499.5 ℓ **iii** 1495 ℓ

b If the actual volume is V litres, complete this statement in each case: $... \leqslant V < ....$

c Explain why there is a 'less than or equal to' sign before the V but a 'less than' sign after the V.

d How do you know that a volume written as 1500 litres has not been measured to the nearest tenth of a litre?

4 Get Real!

ChocoBars should weigh 40 grams with a tolerance of 5% either way.
If the bars weigh 40 grams correct to the nearest 10 grams, will they be within the tolerance? Show how you worked out your answer.
Why do you think that manufacturers have a 'tolerance' in the sizes of their products?

5 Get Real!

What is the maximum possible total weight of 10 cartons, each weighing 1.4 kg correct to the nearest 100 g?
Why might someone need to do a calculation like this in real life?

Explore

- ◉ Write a note to explain to someone else how to find the minimum and maximum possible ages of a person whose age is given as a whole number of years, for example, 8 years

- ◉ Write a note to explain to someone else how to find the minimum and maximum possible amounts of money when the quantity is given to the nearest pound, for example, £18

Investigate further

Rounding

ASSESS

The following exercise tests your understanding of this chapter, with the questions appearing in order of increasing difficulty.

1 4645 people watch Redruth win their latest match.
Write this number

   a to the nearest 1000   **c** to the nearest 50

   b to the nearest 100   **d** to the nearest 10.

2 Copy and complete the table below.

Starting number	To the nearest 10	To the nearest 500	To the nearest 1000
66 329	66 330		
206 021			206 000
211 372		211 500	
332 404			
852 612			

3 The square root of 200 is bigger than one integer (whole number) and smaller than another.

   a What are these two integers?

   b Looking at the square of these two numbers, what is the square root of 200 to the nearest integer?

4 The world's tallest person was Robert Wadlow who died in 1840. He measured 271.78 cm (8 ft 11 in). The shortest adult male was Calvin Phillips who died in 1812. He measured 67.31 cm (2 ft $2\frac{1}{2}$ in).

Write both these metric heights to one decimal place.

5 Round the five numbers in question **2** to one significant figure.

6 Gary says that he and Jane are the same age – to one significant figure. Gary is 44 and Jane is 36. Is Gary correct? Explain your answer.

7 The average person's heart beats about once a second. Estimate how many times it beats during a year.

8 The (movement) energy of an athlete of mass 43 kg running at a velocity of 9.7 m/s can be found by working out $43 \times 9.7 \times 9.7 \div 2$. Use appropriate approximations for 43 and 9.7 and estimate the athlete's energy.

9 Ngugi lives on the equator, which is a circle of diameter 12 756 km. George lives in the UK on a circle of latitude with diameter 7854 km. To calculate the distance each boy moves in one day due to the Earth's rotation we multiply each diameter by 3.14

Write all three values given above correct to **the nearest thousand** and hence estimate how much further Ngugi travels than George in one day.

10 Estimate the value of: **a** $\dfrac{9.6^2 - 4.2^2}{2 \times 9.65}$ **b** $\dfrac{24.8 \times 3.2}{0.54}$

11 a The length of a rectangle is given as 27 m correct to the nearest m. Write down the minimum and maximum possible lengths it could be.

   b A different length is given as 5.0 cm correct to the nearest mm. Write down the minimum and maximum possible lengths it could be.

Try a real past exam question to test your knowledge:

12 a Work out 600×0.3

   b Work out $600 \div 0.3$

   c You are told that $432 \times 21 = 9072$
Write down the value of $9072 \div 2.1$

   d Find an approximate value of $\dfrac{2987}{21 \times 49}$
You **must** show all your working.

Spec A, Int Paper 1, Nov 03

5 Properties of triangles

OBJECTIVES

G **Examiners would normally expect students who get a G grade to be able to:**

Identify isosceles, equilateral and right-angled triangles

Use the word 'congruent' when triangles are identical

E **Examiners would normally expect students who get an E grade also to be able to:**

Show that the angles of a triangle add up to 180° and use this to find angles

Show that the exterior angle of a triangle is equal to the sum of the interior opposite angles

Use angle properties of isosceles, equilateral and right-angled triangles

What you should already know ...

■ Properties of angles at a point and on a straight line

■ Types of angles, including acute, obtuse, reflex and right angles

■ Parallel lines, including opposite angles, corresponding angles and alternate angles

VOCABULARY

Exterior angle – if you make a side of a triangle longer, the angle between this and the next side is an exterior angle

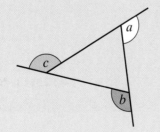

a, b and c are exterior angles

Isosceles triangle – a triangle with 2 equal sides and 2 equal angles; the equal angles are called **base angles**

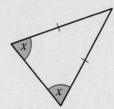

the x angles are base angles

Equilateral triangle – a triangle with 3 equal sides and 3 equal angles – each angle is 60°

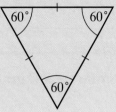

Congruent – exactly the same size and shape; one of the shapes might be rotated or flipped over

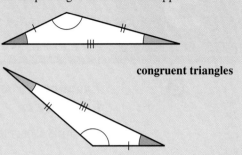

congruent triangles

Right-angled triangle – a triangle with one angle of 90°

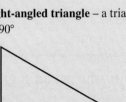

Learn 1   Angle sum of a triangle

Triangle ABC has angles a, b and c.

Suppose side AB is extended to D then BE is drawn from B parallel to AC.

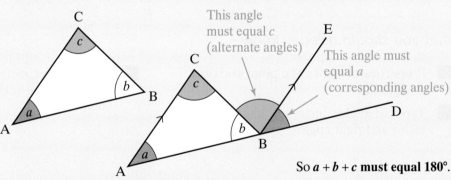

This angle must equal c (alternate angles)

This angle must equal a (corresponding angles)

So $a + b + c$ must equal **180°**.

Example:

In triangle PQR, angle P is 32° and angle Q is 54°.
Calculate the size of angle R and state what type of angle it is.

The sketch shows the triangle PQR.

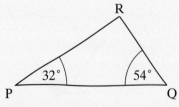

angle P + angle Q = 32° + 54° = 86°

angle R = 180° − 86°

angle R = 94°

The angle R is obtuse.

The sum of the angles of a triangle = 180°

The angle is greater than 90° but less than 180°

Apply 1

1 Work out the angle marked by each letter and state what type of angle it is.

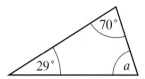 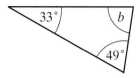 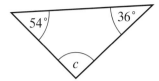

Not drawn accurately

2 Get Real!
The sketch shows the side of a ramp –
it is not drawn accurately.
Calculate the angle marked x.

3 Get Real!
Both sides of a roof are inclined at 34° to the horizontal.
Calculate the angle, x, between the sides of the roof.

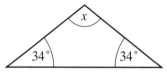

**Not drawn
accurately**

4 In triangle LMN, angle L is 26° and angle M is 108°.
Sketch the triangle and calculate angle N.

5 Calculate the angles marked by letters. Give a reason for each answer.
Choose from: **angles on a straight line**, **angles of a triangle**, **opposite angles.**

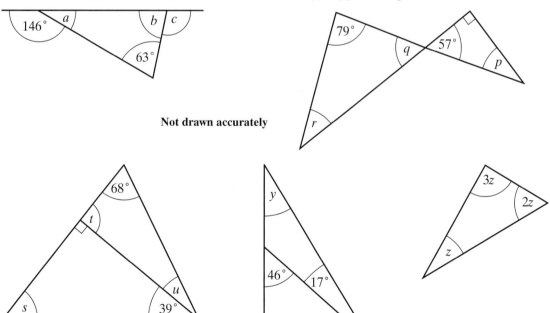

Not drawn accurately

6 In triangle PQR, angle P is 80°. Give five possible pairs of values for
angles Q and R.

7 Calculate the angles marked by letters. Give a reason for each answer.

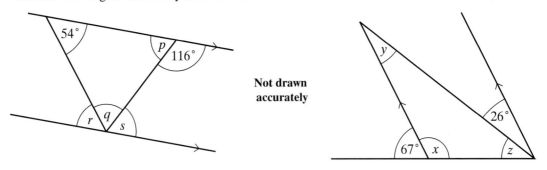

Not drawn accurately

8 Triangle ABC has a right angle at A and angle B equals angle C.
Calculate angle B and angle C.

9 In triangle PQR, angle P is 10° larger than angle R and
angle Q is 20° larger than angle R.
Calculate angles P, Q and R.

Learn 2 Exterior angle of a triangle

When one side of a triangle is extended you get an **exterior** angle.
The exterior angle **is equal to the sum of the interior opposite angles (at the other two vertices)**. Look at the diagram in Learn 1 to see the proof for this.

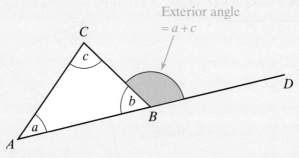

Exterior angle
$= a + c$

Example: Find the angles x and y.

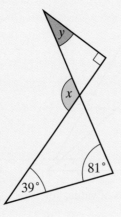

$x = 81° + 39°$ ←——— The angle x is an exterior
$x = 120°$ angle for the lower triangle

$x = y + 90°$ ←——— The angle x is also an exterior
$120° = y + 90°$ angle for the upper triangle
$y = 30°$

Apply 2

1 Work out the angles marked by letters.

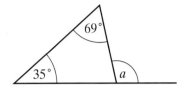

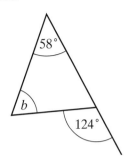

Not drawn accurately

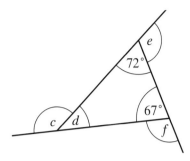

2 Get Real!
The diagram shows the end view of a shelf.
Calculate the angle marked *x*.

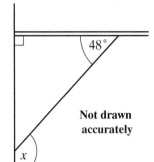

Not drawn accurately

3 Calculate the angles marked by letters. Give a reason for each answer.
Choose from: **exterior angle of a triangle, angles on a straight line**.

Not drawn accurately

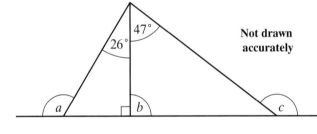

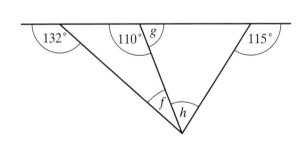

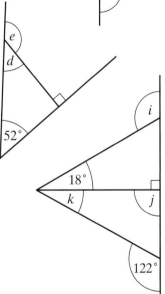

4 One of the exterior angles of a triangle is 115°.
Find two possible sets of values for the interior angles of the triangle.

5 Get Real!
The diagram shows a point P on one
side of a river and two points,
Q and R, on the other side.
The sides of the river are parallel.
Calculate the angles marked *x* and *y*.

Not drawn accurately

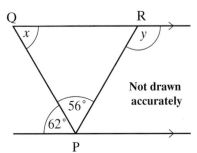

6 Get Real!

The diagram shows the positions of three villages. Charfield is due north of Wickwar, the bearing of Hillesley from Wickwar is 076° and the bearing of Hillesley from Charfield is 120°.

a Calculate the angles marked x and y.

b What is the bearing of

    i Wickwar from Hillesley

    ii Charfield from Hillesley?

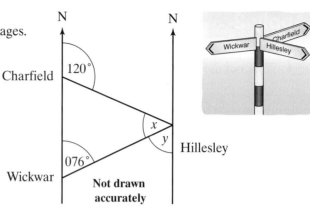

N

Charfield 120°

N

x
y Hillesley

076°

Wickwar

Not drawn accurately

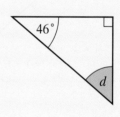
Wickwar — Charfield
Hillesley

Explore

◎ Draw any triangle and extend the sides to make three exterior angles as shown in the diagram

◎ Measure the exterior angles and find their sum

◎ Draw any quadrilateral (4-sided shape) and extend the sides to make four exterior angles

◎ Measure the exterior angles and find their sum

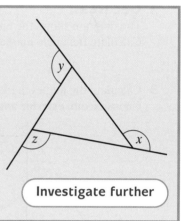

y

z

x

Investigate further

Learn 3   Describing triangles

Examples: Calculate the size of angles a, b, c and d.

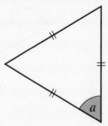

a

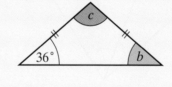

c

36°

b

46°

d

$a = 60°$ Angles of an equilateral triangle
$b = 36°$ Base angles of an isosceles triangle
$c = 108°$ Angles of a triangle add up to 180°
$d = 44°$ Angles of a right-angled triangle

Apply 3

1 Work out the angles marked by letters.

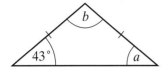

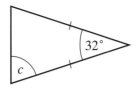

Not drawn accurately

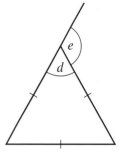

2 Calculate the angles marked by letters.

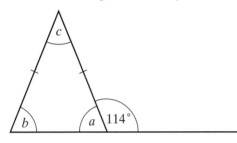

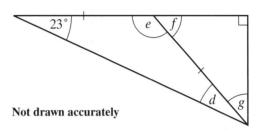

Not drawn accurately

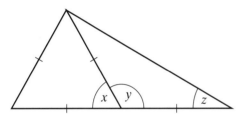

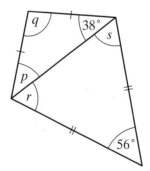

3 Get Real!
The diagram shows the beams in a roof. Calculate the angles *x* and *y*.

Not drawn accurately

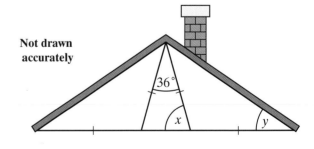

4 Calculate the angles marked by letters. Give a reason for each answer.

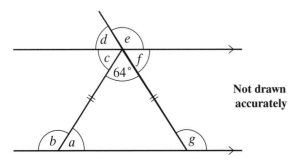

Not drawn accurately

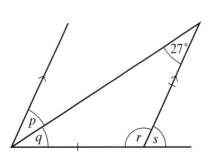

5 Measure the sides and angles of each triangle.
Identify two pairs of congruent triangles.

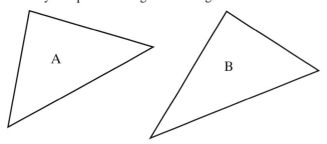

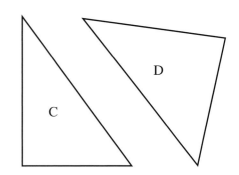

6 One angle of an isosceles triangle is 28°.
Draw two possible triangles and give their other angles.

7 Get Real!
The diagram shows a rectangular gate, ABCD.
The diagonals AC and BD meet at E.

a Identify two pairs of congruent isosceles triangles.

b Identify four congruent right-angled triangles.

c The obtuse angle between the diagonals is 124°.
Calculate the angles x and y.

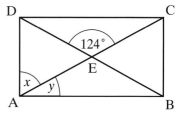

8 a Calculate all the angles in this star.

b How many right-angled triangles are there in the
diagram?

c How many pairs of congruent triangles are there
in the diagram?

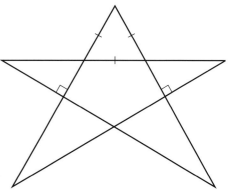

9 a The diagram shows how five congruent
isosceles triangles fit together to make a pentagon.
Calculate the angles, x and y, of one of the isosceles triangles.

b Eight congruent isosceles triangles fit
together in a similar way to make an octagon.
Calculate the angles of one of the isosceles triangles.

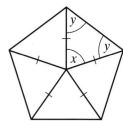

Explore

◎ How many different isosceles triangles can you draw that
have an angle of 30°?

◎ How many different isosceles triangles can you draw that
have an angle of 120°?

Investigate further

Properties of triangles

The following exercise tests your understanding of this chapter, with the questions appearing in order of increasing difficulty.

1 Copy and complete the following statements:

  a A triangle with all sides equal is called ... .

  b A triangle with two sides equal is called ... .

  c A triangle with all angles equal is called ... .

  d A triangle with two angles equal is called ... .

  e A triangle with one angle of 90° is called ... .

  f An equilateral triangle has got all equal ... and ... .

  g An isosceles triangle has got two equal ... and ... .

  h Two triangles that are identical in shape and size are called ... .

2 In each of the triangles below write down the letters of all triangles that are

  a equilateral

  b isosceles

  c right-angled.

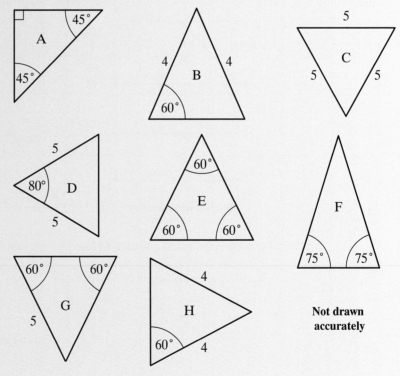

Not drawn accurately

3 Find the values of the angles marked in the diagrams below.

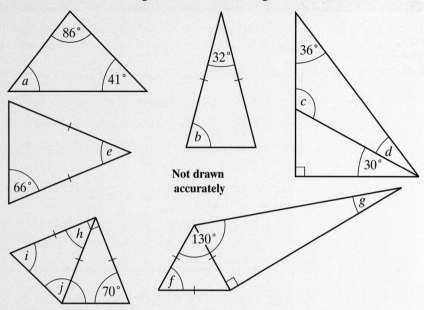

Not drawn accurately

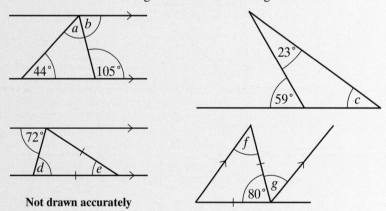

4 An isosceles triangle has a base angle of 65°.
What are the sizes of the other two angles?

5 Find the values of the angles marked in the diagrams below.

Not drawn accurately

Try a real past exam question to test your knowledge:

6 A triangle has angles of 63°, $2x$ and x.

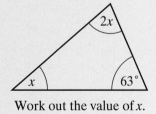

Not drawn accurately

Work out the value of x.

Spec A, Mod 5 Foundation Paper 1, June 03

OBJECTIVES

F ▶ **Examiners would normally expect students who get an F grade to be able to:**

Simplify expressions with one variable such as $a + 2a + 3a$

E ▶ **Examiners would normally expect students who get an E grade also to be able to:**

Simplify expressions with more than one variable such as $2a + 5b + a - 2b$

D ▶ **Examiners would normally expect students who get a D grade also to be able to:**

Multiply out expressions with brackets such as $3(x + 2)$ or $5(x - 2)$

Factorise expressions such as $6a + 8$ and $x^2 - 3x$

C ▶ **Examiners would normally expect students who get a C grade also to be able to:**

Expand (and simplify) harder expressions such as $x(x^2 - 5)$ and $3(x + 2) - 5(2x - 1)$

What you should already know ...

- ▪ Add, subtract and multiply integers
- ▪ Multiply a two-digit number by a single-digit number

VOCABULARY

Variable – a symbol representing a quantity that can take different values such as x, y or z

Term – a number, variable or the product of a number and a variable(s) such as 3, x or $3x$

Algebraic expression – a collection of terms separated by + and − signs such as $x + 2y$ or $a^2 + 2ab + b^2$

Product – the result of multiplying together two (or more) numbers, variables, terms or expressions

Collect like terms – to group together terms of the same variable, for example, $2x + 4x + 3y = 6x + 3y$

Simplify – to make simpler by collecting like terms

Expand – to remove brackets to create an equivalent expression (expanding is the opposite of factorising)

Factorise – to include brackets by taking common factors (factorising is the opposite of expanding)

Linear expression – a combination of terms where the highest power of the variable is 1

Linear expressions	Non-linear expressions
x	x^2
$x + 2$	$\frac{1}{x}$
$3x + 2$	$3x^2 + 2$
$3x + 4y$	$(x + 1)(x + 2)$
$2a + 3b + 4c + ...$	x^3

Learn 1 Collecting like terms

Examples: Simplify the following expressions.

a $3f + 2g + 4f - 3g + 3 = 7f - g + 3$ ⟵ $3f + 4f = 7f$

b $4ad + 3a^2 + 2da + a^2 = 6ad + 4a^2$ ⟵ $+3a^2 + a^2 = 4a^2$

⟵ $4ad + 2da = 4ad + 2ad = 6ad$
Note: $2da$ is the same as $2ad$

Apply 1

1 Simplify the following by collecting like terms.

a $a + a + a + a + a + a$ **g** $2k + 4k + k + 2k$ **m** $y - y - y + y + y$

b $p + p + p + p$ **h** $2m + 3m + 2m + m$ **n** $2d + 5d - d - 3d + 2d$

c $k + k$ **i** $5q + 2q + 2q + q$ **o** $4a - 6a$

d $d + d + 2d + d$ **j** $t + t + t + t + t - t$ **p** $a + a - a - a$

e $3y + 2y$ **k** $g + g + g + g + g - g - g - g$ **q** $3p + 4p - 2p$

f $2m + m + 4m$ **l** $x + x - x + x - x$ **r** $-2a + 4a$

2 Michelle writes the answer to $p + p + p + p + p$ as p^5. Kwame writes the answer as $p5$. Is anyone correct?

3 Toby thinks the answer to $4a + 2a + 3d + d = 10ad$. Is he correct?
Explain your answer.

4 Simplify the following by collecting like terms.

a $x + 2x + 3 + x$ **f** $a + b + 2a$ **k** $5y + 6 - 4y - 6$

b $y + y + 5 + y + 3$ **g** $p + 2q + 3p + 4q$ **l** $4y - 6y - 2$

c $2f + 6 + 3f - 2$ **h** $3t + 6g + 4t - 2g$ **m** $6p - 4 - 2p - 3 - 5p$

d $7y + 2y + 1$ **i** $2f + 3g + 4 - 2g - f$ **n** $2w + 4t - 3t - w$

e $3x - 2x + 6 + 4x + 2$ **j** $2p + 6 - p - 2 - p$

5 The answer is $5q$. Write down five questions that have this answer.

6 Get Real!

The diagram shows an L-shaped floor with dimensions as shown.
A carpenter is trying to work out the length of wood needed to make the
skirting boards. Find an expression for the perimeter of the room in
terms of x.

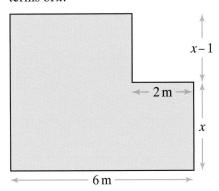

7 Louise is asked to simplify $4ab + 5ba$. She thinks it is impossible.
Is she correct?

8 Simplify the following by collecting like terms.

a $4ab + 2ab$

b $3pq + 2pq$

c $2ef + ef + 5ef$

d $3cd + 2cd + 4dc$

e $6gh + 2hg + gh$

f $6mn + mn - 3nm$

g $2baa + 3baa$

h $6tom + 2mot + 4omt$

i $2xy + 3xy - xy$

j $8mn + 2mn - 3mn$

k $7pq + 3qp - 2pq$

l $8bc - 2bc + 3cb$

m $3pqr + 2prq + 3qpr - rpq$

n $2ab + a + b$

o $7gh + 3g - 2g + hg$

9 The answer is $7ab + 2cd$. Write down five questions that have this answer.

10 $x^2 + x^2 + x^2 = 6x$

Do you agree with this statement? Give reasons for your answer.

11 Simplify the following by collecting like terms.

a $p^2 + p^2$

b $y^3 + 2y^3 + 3y^3$

c $4t^5 + 2t^5 + t^5$

d $5t^6 + 2t^6 - 3t^6$

e $3p^2 + p + 2p^2$

f $4a^2 + a^3 + 2a^2 + 3a^3$

g $4y^3 + 2y^2 + 3y^3 - y^2$

h $3d^2 + 2h^3 + d^3 - 4h^2$

i $5f^4 + g^6 - 2f^4 + 2g^6$

j $4x^3 + 2y^2 + x^2 + 3x^3$

k $2xy^2 + 3xy^2$

l $6pq^3 + 2pq^3 + 3q^3p$

m $4ab^2 + 2ab^2 - 3ab^2$

n $4gh^2 - 2hg^2 + gh^2$

12

| **A** $2x + 3xy$ | **B** $4x + 6yx$ | **C** $2x^2 + 3xy^2$ | **D** $2x + 3x^2y$ |

Simplify the following by collecting like terms.

a $\mathbf{A} + \mathbf{B}$

b $\mathbf{B} + \mathbf{A}$

c $\mathbf{A} - \mathbf{B}$

d $\mathbf{A} + \mathbf{B} + \mathbf{C}$

e $\mathbf{A} + \mathbf{B} + \mathbf{C} + \mathbf{D}$

f $\mathbf{C} + \mathbf{D}$

g $\mathbf{B} + \mathbf{D}$

h Can you spot a link between **A** and **B**?

13 a The answer is $12p^2$. Find five questions with this answer.

b The answer is $12p^2 + 6q^3$. Find five questions with this answer.

c The answer is $12p^2q + 6pq^2$. Find five questions with this answer.

d The answer is $12p^2q + 6p + 6pq^2$. Find five questions with this answer.

Explore

The grid below is part of a 10×10 number square

41	42	43	44	45
31	32	33	34	35
21	22	23	24	25
11	12	13	14	15
1	2	3	4	5

The shaded cells form an L-shape

◎ Add together the four shaded cells and write down the total

◎ Multiply the bottom number (3) by 4 and add 31

◎ Is the total the same?

◎ Try it with a different L

> **Investigate further**

Learn 2 Expanding brackets

Examples: **a** Expand $5(2y - 1)$.

	2y	**−1**
5	10y	−5

$5 \times -1 = -5$

$5(2y - 1) = \boxed{10y - 5}$ ← Write $10y - 5$ not $10y + -5$

$10y - 5$ is a linear expression because the highest power of the variable (y) is 1.

b Expand $p(p^2 - 5)$.

	p²	**−5**
p	p³	−5p

$p \times -5 = -5p$

$p(p^2 - 5) = \boxed{p^3} - 5p$

$p \times p^2 = p \times p \times p = p^3$

Apply 2

1 Complete the following:

a $3(t + 4) = 3t + \dots$

b $5(p + 2) = 5p + \dots$

c $6(h + 3) = \dots + 18$

d $4(2a + 3) = 8a + \dots$

e $3(3y + 2) = 9y + \dots$

f $5(y - 2) = 5y - \dots$

g $4(b - 3) = 4b - \dots$

h $7(d - 2) = \dots - 14$

i $8(2d - 3) = 16d - \dots$

j $5(4p - 4) = \dots - 20$

k $-2(a + 6) = -2a \dots$

l $-3(b - 4) = -3b \dots$

2 Multiply these out:

a $4(x + 2)$

b $6(y + 3)$

c $3(m + 2)$

d $5(b + 3)$

e $6(t + 4)$

f $3(2y + 3)$

g $5(3p + 6)$

h $\frac{1}{3}(6x - 15)$

i $\frac{1}{4}(16f - 4)$

j $7(g - 4)$

k $\dfrac{35h + 10}{5}$

l $-4(q + 2)$

m $-2(a + 3)$

n $-5(2m - 3)$

o $-7(-4a - 1)$

p $4(h - 1)$

q $\frac{1}{2}(4b + 6)$

3 Sam thinks the answer to $5(3x - 2)$ is $15x - 2$. Hannah says he is wrong.
Who is correct and why?

4 Complete the following:

a $x(x + 2) = x^2 + \ldots$

b $p(p + 4) = \ldots + 4p$

c $x(x - 2) = x^2 \ldots$

d $f(f - 3) = \ldots - 3f$

e $y(y^2 - 3) = \ldots - 3y$

f $2f(f - 2) = \ldots$

5 Expand:

a $p(p + 3)$

b $b(b - 4)$

c $a(5 + a)$

d $x(x^2 + 3)$

e $w(ig + am)$

f $x(x^3 + 4x)$

g $t(t^2 - 1)$

h $mu(fc - mu)$

i $h^2(h^3 + 4)$

j $p(p^2 - 7)$

6 The answer is $12y - 36$.
Write down five questions of the form $a(by + c)$ with this answer.
(a, b, and c are integers – positive or negative numbers.)

Explore

◎ Think of a number

◎ Add 2

◎ Multiply the new total by 4

◎ Halve your answer

◎ Subtract twice the original number

◎ The answer is 4

Investigate further

Learn 3 Expanding brackets and collecting terms

Example: Expand and simplify $3(x - 2) - 5(2x - 1)$.

Treat this as two separate algebraic expressions, $3(x - 2)$ and $-5(2x - 1)$, and merge the answers together at the end

Step 1 Expand $3(x - 2)$.

	x	-2
3	$3x$	-6

$3(x - 2) = 3x - 6$

Step 2 Expand $-5(2x - 1)$.

	$2x$	-1
-5	$-10x$	$+5$

$\leftarrow\ -5 \times -1 = +5$

$-5(2x - 1) = -10x + 5$

Step 3 Merge the two answers by collecting like terms.

$3(x - 2) - 5(2x - 1) = 3x \enclose{circle}{-6} - 10x \enclose{circle}{+5}$
$= -7x - 1$

Underlining or circling like terms (including their sign) helps when collecting them: $3x - 10x = -7x$ and $-6 + 5 = -1$

Apply 3

1 Complete the following.

a $2(y + 3) + 3(y + 2)$
$= 2y + 6 + \ldots + \ldots$
$= \ldots y + \ldots$

b $3(d + 2) + 4(d - 3)$
$= \ldots + \ldots + 4d - \ldots$
$= \ldots d - \ldots$

c $5(2x + 2) - 3(2x - 3)$
$= 10x + \ldots - \ldots x + \ldots$
$= \ldots x + \ldots$

2 Expand and simplify:

a $4(x + 2) + 2(x + 3)$

b $2(p + 3) + 3(2p - 4)$

c $6x - (2 - x)$

<u>**d**</u> $3(m - 1) - 4(m - 2)$

<u>**e**</u> $\frac{1}{2}(6y - 3) + \frac{1}{4}(12 - 4y)$

<u>**f**</u> $4x - (x + 2)$

g $4(t - 2) - 2(t + 1)$

3 Simplify:

a $4(2m - 3) + 3(m - 6)$

b $3(2a - 1) - 3(4 - a)$

<u>**c**</u> $5(6x - 3) + 2(3 - 2x)$

d $4(2y - 1) - 4(3y - 5)$

e $5(2t - 4) - 7(2 - 3t)$

<u>**f**</u> $2x(3y + 1) + 3y(2x - 1)$

4 Josie thinks the answer to $3(2m - 1) - 4(m - 2)$ is $2m - 11$.
Explain what she has done wrong.

<u>**5**</u> Find the integers a and b if $4(x - a) - b(x - 1) = 2x - 14$.

6

| **A** $3y + 2$ | **B** $2y - 3$ | **C** $5y - 1$ | **D** $y + 2$ |

Expand and simplify:

a $2\mathbf{A} + 3\mathbf{C}$

b $\mathbf{C} - 2\mathbf{D}$

c $2\mathbf{A} - \mathbf{B}$

<u>**d**</u> Work out a combination of two cards that gives the answer 13.

7 Get Real!

The diagram shows an L-shaped floor with dimensions as shown.
The floor is to be covered with tiles, each measuring 1 m by 1 m.

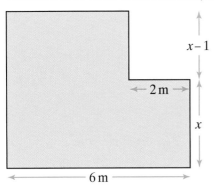

a By splitting the floor into two rectangles, calculate the area of the floor.

b By splitting the floor into two different rectangles, calculate the area of the floor.

c Are your answers the same each time?
Give a reason for your answer.

Explore

◎ Pick a blue card

◎ Double the number

◎ Add 1 to the new number

◎ Multiply the new number by 5

◎ Pick a white card

◎ Add this number to the previous result

◎ Subtract 5

◎ What do you notice?

| 5 | 6 | 7 | 8 |
| 1 | 2 | 3 | 4 |

Investigate further

Learn 4 Factorising expressions

Examples:

a Factorise $5a - 10$

$5a - 10 = 5\,(a - 2)$

$5a = \textcircled{5} \times a$ and $10 = 2 \times \textcircled{5}$
Both terms have 5 as a common factor

b Factorise $x^2 - 3x$

$x^2 - 3x = x\,(x - 3)$

$x^2 = x \times \textcircled{x}$ and $3x = 3 \times \textcircled{x}$
Both terms have x as a common factor

Treat factorising as the 'reverse' of expanding.

HINT You can always check if you have factorised correctly by multiplying the bracket out to make sure you get the question.

Apply 4

1 Complete the following.

   a $2y + 10 = 2(y + ...)$ **d** $18xy - 30ab = 6(3xy - ...)$

   b $4g + 20 = ... \,(... + 5)$ **e** $20ab + 35\,cd - 10ef = 5(4ab + ...\,cd - ...)$

   c $12d + 42 = 6(... + ...)$

2 Factorise the following expressions.

   a $2x + 6$ **g** $21 - 42t$ **m** $18f - 36g + 12h$

   b $3y + 12$ **h** $44t - 55c$ **n** $20xy + 60tu - 80pq$

   c $7y - 63$ **i** $13a + 23b$ **o** $2f - 11g + 33h$

   d $8y - 40$ **j** $20b + 35a + 15c$

   e $14y + 20$ **k** $-12a - 16$

   f $6t - 18$ **l** $-42ab + 70cd + 14ef$

> **HINT** Two of these expressions cannot be factorised.

3 Sandra thinks that $12a + 18$ factorised completely is $3(4a + 6)$.
Is she correct? Explain your answer.

4 The answer is $4(...\,p + ...)$ where $...\,p + ...$ is an expression.
Find five expressions that can be factorised completely with answers in this form.

5 Rashid thinks he has factorised $2x + 3y$ correctly as $2(x + 1\frac{1}{2}y)$.
Rana thinks the expression cannot be factorised? Who is correct and why?

6 Complete the following.

   a $4cd + 7c = c(4d + ...)$ **c** $x^2 + 5x = x(x + ...)$

   b $4ab + 9bc = b(... + 9c)$ **d** $5abc - 8bcd = bc(5a - ...)$

7 Factorise the following expressions.

   a $2ab + 5b$ **f** $x^3 - 3x$ **k** $wig - wam$

   b $3cd + 7c$ **g** $a^2 - a$ **l** $bugs + bunny$

   c $2pq + p$ **h** $6xy + 7xyz$ **m** $hocus - pocus$

   d $3xy + 5xz$ **i** $3de + 5fg$ **n** $silly + billy$

   e $p^2 + 2p$ **j** $x^2 - 3x$ **o** $wonkey - donkey$

8 Copy these two tables.
Match the expression with the correct factorisation buddy.
Fill in the missing buddies.

Expression
$2x^2 + 8x$
$6x^2y - 3xy$

Factorisation
$2x(x + 4)$
$3xy(2y - 1)$
$x(x + 8)$

Explore

◎ Pick five consecutive numbers

◎ Add them together

◎ Is the answer a multiple of 5?

◎ Investigate further by picking five different consecutive numbers

Investigate further

Use of symbols

ASSESS

The following exercise tests your understanding of this chapter, with the questions appearing in order of increasing difficulty.

1 Simplify the following expressions.

a $4a + 7a + 9a$ **d** $3f - 4f - f$ **f** $3k - 4k + 5k - 6k + 8k + k - 4k - 2k$

b $5b - 9b + 6b$ **e** $4g + 6g - 10g$ **g** $2a + 3a + 4$

c $4c - 7c$

2 Simplify the following expressions.

a $2m + 3n + 4m + n$ **f** $a - b - a + b + c$

b $9v - 3w + 6v - 5w$ **g** $6a + 3m - 2y - 5a + 3y - 2m$

c $5e - 3u - 2u - e$ **h** $q + u + e + u + e$

d $4x + 4y - 4x - 3y$ **i** $m + i + s + s + i + s + s + i + p + p + i$

e $w + z - w$

3 a Simplify the following expressions where possible.

   i $4w + 3w^2 - w^2 - 5w$ **iv** $3bil + 2ben - 2lbi - ben$

   ii $2 - 3x + 4x^2 + 9x^3 - 5 + 4x - 4x^2 + 2x^3$ **v** $7t + 9u - 6u - t^2$

   iii $3de + 5ef$

b Remove the brackets from the following expressions.

   i $2(a + 3)$ **v** $7(a + 2b)$ **ix** $-6(-4a - 3b)$

   ii $5(3a - 1)$ **vi** $5(6a - 3b)$ **x** $-2(a - 2b + 3)$

   iii $-3(4a + 5)$ **vii** $-3(3a + 2b)$

   iv $-7(3a - 6)$ **viii** $-7(-3a - 2b)$

4 Factorise the following expressions.

a $2a + 10$ **f** $10j + 15k - 20l$ **k** $4a^2 - 6a$

b $10b - 12$ **g** $30p - 45q - 75r$ **l** $bil + ben$

c $16 - 4c$ **h** $5x + 15y$ **m** $abra + cadabra$

d $5d + 20e + 35f$ **i** $2ab - 3a$ **n** $12cat + 3sat - 6mat$

e $6g - 9h + 12i$ **j** $x^2 + 7x$

5 Remove the brackets from the following expressions and simplify.

a $2c + 3(c + 5)$ **h** $5(2c - 3) - 3(-2c - 1)$

b $2c - 3(c + 5)$ **i** $3(2c - d + 3e) + (5c + 3d - 2e)$

c $2c + 3(c - 5)$ **j** $3(2c - d + 3e) - (5c + 3d - 2e)$

d $2c - 3(c - 5)$ **k** $y(y^2 - 7)$

e $4(c - 3) + 2(c + 7)$ **l** $2z(5z^2 + 4z - 8)$

f $4(c - 3) - 2(c + 7)$ **m** $x(x - 3) + 5(x - 3)$

g $5(2c - 3) - 3(2c + 1)$ **n** $p(p^2 + 3p - 4) - 6(p^2 + 3p - 4)$

6 a Simplify $2x + 3y + 5x - 2y - 4x$

b Factorise $4c + 12$

c Factorise $x^2 + 5x$

7 a Simplify $5p + 2q - q + 2p$

b Multiply out $4(r - 3)$

c Multiply out $s(s^2 + 6)$

d Simplify fully $(2t^3u) \times (3tu^2)$

7 Decimals

OBJECTIVES

F → **Examiners would normally expect students who get an F grade to be able to:**

Write down the place value of a digit such as the value of the 4 in 0.24

Order decimals, for example, which is bigger, 0.24 or 0.3?

E → **Examiners would normally expect students who get an E grade also to be able to:**

Add and subtract decimals

D → **Examiners would normally expect students who get a D grade also to be able to:**

Multiply two decimals such as 2.4 × 0.7

Convert decimals to fractions and fractions to decimals

C → **Examiners would normally expect students who get a C grade also to be able to:**

Divide a number by a decimal such as 1 ÷ 0.2 and 2.8 ÷ 0.07

What you should already know ...

■ Add, subtract, multiply and divide whole numbers

VOCABULARY

Digit – any of the numerals from 0 to 9

Integer – any positive or negative whole number or zero, for example, $-2, -1, 0, 1, 2, ...$

Decimal – a number in which a decimal point separates the whole number part from the decimal part, for example, 24.8

Numerator – the number on the top of a fraction

$$\text{Numerator} \longrightarrow \frac{3}{8} \longleftarrow \text{Denominator}$$

Denominator – the number on the bottom of a fraction

Terminating decimal – a decimal that ends, for example, 0.3, 0.33 or 0.3333

Recurring decimal – a decimal with a repeating digit or group of digits, for example, 0.33333333333 ... (written as $0.\dot{3}$) or 0.25678678678678 ... (written as $0.25\dot{6}7\dot{8}$)

Learn 1   Place value

The decimal point separates the whole numbers from the fractions.

12	.	345
Whole numbers	.	Fractions

Examples:

a What is the value of the 2 in 0.425?

Write out the number with the headings.

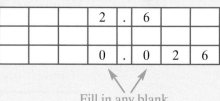

Thousands	Hundreds	Tens	Units		Tenths	Hundredths	Thousandths
			0	.	4	2	5

The value of 2 is 2 hundredths.

b Calculate 2.6 ÷ 100.

Remember the digits move one place for each 0 – one place for 10, two for 100, three for 1000, ...

			2	.	6		
			0	.	0	2	6

Fill in any blank spaces with zeros

2.6 ÷ 100 = 0.026

Apply 1

1 Put these numbers in order of size, starting with the smallest.

   2.4   3   2.06   2.175   1.999   0.987

2 Write down the value of the digit 5 in each of these numbers.

   a 3.5   **c** 5.34   **e** 1.235   **g** 2.514

   b 2.45   **d** 0.156   **f** 523.46   **h** 5432.1

3 Bronwyn says that 2.4 × 10 is 2.40.
   What has she done wrong?

4 Write down the answers to these questions.

   a 2.4 × 10   **c** 5 ÷ 10   **e** 4.5 × 100   **g** 4.2 × 100 ÷ 10

   b 4.2 × 1000   **d** 3.2 ÷ 1000   **f** 0.02 ÷ 1000   **h** 2 ÷ 100 × 10

5 How many lots of 0.2 make 200?

6 Get Real!

The Healthy Bite Café serves 1000 drinks of orange juice every day. The staff serve the drinks in cups containing 0.3 litres of orange juice. They charge £0.85 for each cup.

a How many litres of orange juice do they sell in a day?

b How much money do they take for the sales of orange juice?

7 The grid below gives correct calculations from left to right and top to bottom. (Every blue arrow shows a correct calculation).

For example, $2 \div 10 = 0.2$

For example, $2 \times 10 = 20$

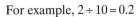

2	÷ 10	0.2	× 1000	200
× 10		÷ 10		÷ 100
20	÷ 1000	0.02	× 100	2
÷ 100		× 10 000		× 10
0.2	× 1000	200	÷ 10	20

Copy these grids and fill in the gaps in the same way.

4.2	÷ 10		× 100	
× 10		÷ 10		÷ 100
	÷ 1000		× 10	
÷ 100		× 100		× 100
	× 10		× 10	

4.3		43		0.43
430		4.3		43
4.3		43		430

8 £1 is worth $1.84

a How many dollars is £10 worth?

b How many dollars is £1000 worth?

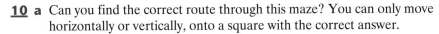

9 Niranjan puts a number into his calculator. He then performs these calculations:

$\times 100 \div 10 \div 100 \times 1000 \div 10$

His calculator gives an answer of 23.4 – with what number did he start?

10 a Can you find the correct route through this maze? You can only move horizontally or vertically, onto a square with the correct answer.

Start	3.4×10 = 34	2.7×100 = 2.700	$30.8 \div 100$ = 0.308	$51.2 - 10$ = 41.2	$1.09 \div 100$ = 0.0109
3.4×10 = 0.34	$42 \div 100$ = 0.42	$12.4 - 10$ = 2.4	$0.2 \div 100$ = 0.002	$14.5 \div 10$ = 0.145	$3 \div 1000$ = 0.003
2.1×100 = 21	2.7×10 = 20.7	$4 \div 100$ = 0.40	$3 \div 10$ = 30	0.7×100 = 700	$0.2 \div 100$ = 0.002
0.1×100 = 10	$20 \div 100$ = 0.2	$40.9 \div 10$ = 4.09	$20.6 \div 10$ = 2.6	0.4×100 = 40	0.1×100 = 10
$1.4 + 1$ = 2.4	$3.2 + 100$ = 320	8.03×10 = 80.3	6.12×100 = 621	1.9×1000 = 1900	0.12×10 = 0.120
End	$15.3 - 10$ = 1.53	$2 \div 1000$ = 0.002	1.4×10 = 14	$0.9 \div 100$ = 0.009	$3.7 \div 10$ = 0.037

b Now correct the wrong answers.

Explore

Kate says that if you start with any number and run it through this machine chain, you get back to your starting number ...

×10   ÷100   ×1000   ÷100

Try it and see.

◎ Make up some machine chains of your own that take you back to your starting number – try to find at least two different chains

◎ Make up some chains of different lengths which take you back to the starting number

◎ Make up some chains that do not return to the starting value
Can you predict the effect?

Investigate further

Learn 2 Adding and subtracting decimals

Examples: **a** 4.2 + 5.1   **b** 6.1 – 2.8   **c** 5.84 + 1.4   **d** 3.29 – 1.4   **e** 5.2 – 2.46

Adding and subtracting decimals is very much like adding and subtracting whole numbers. Just line up the decimal points to make sure you are adding or subtracting digits with the same place value

$$
\begin{array}{r}
4.2 \\
+5.1 \\
\hline
9.3
\end{array}
\qquad
\begin{array}{r}
{}^5\!\!\not{6}.{}^1\!1 \\
-2.8 \\
\hline
3.3
\end{array}
\qquad
\begin{array}{r}
5.84 \\
+1.40 \\
\hline
{}_17.24
\end{array}
\qquad
\begin{array}{r}
{}^2\!\!\not{3}.{}^12\,9 \\
-1.40 \\
\hline
1.89
\end{array}
\qquad
\begin{array}{r}
{}^4\!\!\not{5}.{}^{11}2\,{}^10 \\
-2.46 \\
\hline
2.74
\end{array}
$$

To avoid mistakes, put 0 in any 'spaces' to make both numbers line up on the right

Apply 2

1 Work these out:

a 4.2 + 3.7

b 1.2 + 8.8

c 4.7 + 6.6

d 8.23 + 3.56

e 12.1 + 4.9

f 2.1 + 4.32

g 6.32 + 4.1

h 4.3 + 5.97

i 2.345 + 3.456

j 2.456 + 4.32

k 6.4 + 7.734

l 3 + 4.2 + 5.46

2 Yusef says that 3.2 + 1.34 = 4.36.
Isaac says 3.2 + 1.34 = 4.54.
Who is correct?
Give a reason for your answer.

3 Work these out:

a 4.5 – 3.2

b 6.2 – 1.2

c 5.7 – 3.9

d 7.43 – 3.39

e 8.1 – 5.9

f 8.24 – 4.2

g 7.3 – 3.18

h 9.3 – 5.86

i 7.385 – 4.727

j 8.345 – 5.61

k 3.8 – 2.199

l 4.7 – 1.345

4 Bill says that 4.6 – 1.32 = 3.28
Ted says 4.6 – 1.32 = 3.32
Who made the mistake?
Give a reason for your answer.

5 Work these out:

a 4.2 + 5.1 – 2.8

b 3.1 – 1.7 + 5.2

c 1.9 + 2.42 – 1.6

d 2.15 – 1.4 + 3.2

e 2.4 – 1.23 + 4.56 – 1.2

f 6.1 – 4.56 – 1.2

g 2.76 – 1.4 – 1.23

h 9 – 2.34 – 1.8

i 21.2 – 14.6 + 1.3 – 2.45

6 Fill in the missing numbers (shown as ☺) in these calculations.

a 2.4 + 1.3 = ☺.7

b 5.4 – 1.2 = ☺.☺

c 3.☺ + 2.4 = 5.8

d ☺.7 – 2.8 = 2.☺

e ☺.2 + 4.☺ = 7.1

f 9.☺ – 4.2 = ☺.5

7 This question is about the four numbers below.

A	B	C	D
2.4	4.62	6.9	9.16

a Work out **C** + **D**.

b Work out (**A** + **D**) – (**B** + **C**).

c Work out **B** – **A**, **C** – **B**, and **D** – **C**. Use your answers to decide which two numbers are closest together.

d Add your three answers to part **c** together.

e Calculate **D** – **A**. What do you notice?

8 Get Real!

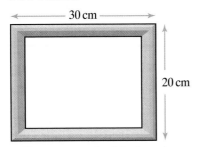

30 cm

20 cm

A picture frame measures 30 cm by 20 cm.

Oliver drops it, and it breaks into eight pieces.
Here are the measurements of each piece.

4.6 cm	8.6 cm
12.6 cm	15.4 cm
8.6 cm	11.4 cm
17.4 cm	21.4 cm

a Can you put the bits in pairs to rebuild the frame? (You need to make two lengths of 20 cm and two lengths of 30 cm.)

b Can you use the same pieces to make a frame that measures 24 cm by 26 cm instead?

9 The questions below all have the same four numbers in them.

a Which three of these arrangements have the same answer?

    i $2.4 + 3.73 - 1.6 - 3.2$ **iii** $3.73 - 1.6 + 3.2 - 2.4$ **v** $3.73 + 2.4 - 3.2 - 1.6$

    ii $3.73 - 1.6 + 2.4 - 3.2$ **iv** $3.73 + 1.6 - 2.4 - 3.2$ **vi** $3.2 + 2.4 - 3.73 - 1.6$

b Why do they have the same answer?

c Can you use what you have discovered to make these questions easier
by avoiding going into negative numbers?

    i $4.1 - 5.6 + 3.8$ **iii** $2 - 4.8 + 3.2$

    ii $1.2 - 5 + 6.4$ **iv** $1.2 - 4.56 - 3.1 + 10$

10 Imagine you have three bricks like this one:

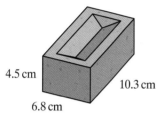

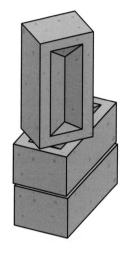

4.5 cm   10.3 cm

6.8 cm

You can put them in a line, you can stack them up,
you can turn them round, ...
What different heights can you make using one, two or all three bricks?

Explore

◎   Find two numbers that add up to 9.4 but have a difference of 1.2

◎   Find two numbers that add up to 5.3 but have a difference of 1.5

◎   Find two numbers that add up to 6.2 but have a difference of 3.1

Is there a quick way to find the numbers?

      Investigate further

Explore

Harry has stamps costing £0.27 each, and stamps costing £0.19 each

◎   Show how he could use these stamps to post a parcel costing £1.57

◎   Which amounts between £1 and £2 can he make exactly?

      Investigate further

Learn 3 Multiplying decimals

Example:

Calculate 0.78×5.2

First remove decimal points: 78×52

Then multiply in your usual way
(The grid method is shown here,
but use your usual method.)

×	**70**	**8**
50	3500	400
2	140	16

$$\begin{array}{r} 3500 \\ 400 \\ 140 \\ +\ \ 16 \\ \hline 4056 \end{array}$$

Finally, put the decimal point back in the answer.

Estimate that 0.78×5.2 is about $1 \times 5 = 5$.

So $0.78 \times 5.2 = 4.056$

Alternatively, count up the number of decimal places in the question.

There are three decimal places in the question: 0.78×5.2

So you need three decimal places in the answer: 4.056

So $0.78 \times 5.2 = 4.056$

Apply 3

1 Use the multiplication $23 \times 52 = 1196$ to help you to complete the questions.

  a 2.3×52 **d** 0.23×52 **g** 0.023×0.052

  b 2.3×5.2 **e** 0.23×0.52

  c 0.23×5.2 **f** 0.23×0.052

2 Calculate:

  a 0.13×22 **e** 1.7×0.22 **i** 8.7×2.5

  b 1.5×2.3 **f** 3.2×13 **j** 8.9×0.16

  c 0.7×1.3 **g** 5.1×2.3 **k** 73.1×0.12

  d 1.1×4.5 **h** 2.7×0.13 **l** 14.3×2.3

3 Now check your answers in question **2** are correct by estimating.

4 Using your answers to question **2**, write down the answers to these.

  a 1.3×22 **e** 17×2.2 **i** 0.087×0.025

  b 1.5×0.23 **f** 0.032×13 **j** 0.0089×0.016

  c 0.07×1.3 **g** 0.0051×0.23 **k** 7.31×1.2

  d 0.11×0.45 **h** 27×1.3 **l** 1430×23

5 A can of Fizzicola contains 0.3 litres of drink. A box holds 36 cans. How many litres of Fizzicola are there in a box?

6 Here are two multiplagons. On each straight line, the numbers in the circles multiply together to make the number in the rectangle. Your job is to copy and complete the multiplagons by filling in the missing numbers.

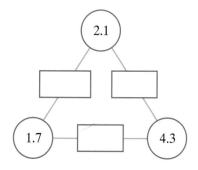 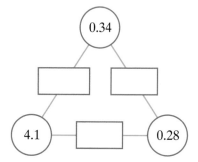

7 Get Real!

Rachel is making some curtains. She buys 4.2 metres of fabric. The fabric costs £3.80 per metre.

a How much does Rachel have to pay?

b The fabric is 1.2 metres wide. What area of fabric has Rachel bought?

8 Toby says $0.4 \times 0.2 = 0.8$
Austin says it isn't, because $4 \times 0.2 = 0.8$
Austin says $0.4 \times 0.2 = 0.08$
Toby says it isn't because that's less than you started with.
Who is right, Toby or Austin?
Give a reason for your answer.

9 a You know $3 \times 2 = 6$.
So what is 0.3×0.2?

b What other multiplications have the same answer as 0.3×0.2?

c Write down five multiplications with an answer of 0.12

10 a Work out the area of the shape on the right.

b Estimate the area of the shape to make sure your answer is the right size.

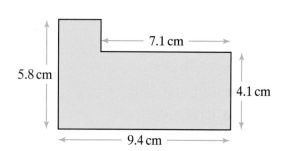

Explore

- Add together 1.125 and 9
- Now multiply 1.125 by 9
- You should get the same answer to both questions

Can you find other pairs of numbers with this characteristic?

Can you find a pair where the product is twice the sum?

Investigate further

Learn 4 Dividing decimals

Examples:

a Calculate $31.2 \div 0.4$

$$31.2 \div 0.4 = \frac{31.2}{0.4}$$

$$= \frac{31.2}{0.4} \overset{\times 10}{\underset{\times 10}{=}} \frac{312}{4}$$

Make the fraction an equivalent fraction by multiplying the numerator and denominator by 10

$$\begin{array}{r} 78 \\ 4\overline{)312} \end{array}$$ ← Now do the division

So $31.2 \div 0.4 = 78$

b Calculate $3.8 \div 0.05$

$$3.8 \div 0.05 = \frac{3.8}{0.05}$$

$$= \frac{3.8}{0.05} \overset{\times 10}{\underset{\times 10}{=}} \frac{38}{0.5} \overset{\times 10}{\underset{\times 10}{=}} \frac{380}{5}$$

Make the fraction an equivalent fraction by multiplying the numerator and denominator by 10 and 10 again

$$\begin{array}{r} 76 \\ 5\overline{)380} \end{array}$$ ← Now do the division

So $3.8 \div 0.05 = 76$

Apply 4

1 Write these calculations as equivalent fractions and work them out.

a $3.2 \div 0.4$ e $53.1 \div 0.3$ i $0.056 \div 0.7$

b $25.4 \div 0.2$ f $1.74 \div 0.6$ j $1.32 \div 0.004$

c $2.85 \div 0.5$ g $0.4 \div 0.08$ k $0.028 \div 0.7$

d $42.2 \div 0.02$ h $32 \div 0.8$

2 Write these calculations as equivalent fractions and work them out.

a $4.07 \div 1.1$ e $16.8 \div 0.12$ i $0.0552 \div 0.012$

b $22.8 \div 1.2$ f $25.3 \div 0.11$ j $0.945 \div 1.5$

c $2.73 \div 0.13$ g $7.392 \div 0.11$ k $2.46 \div 0.0015$

d $0.264 \div 1.1$ h $0.474 \div 0.12$ l $56.2 \div 0.25$

3 Get Real!

Malcolm the plumber has a 6 metre length of copper pipe.
He needs to cut it into 0.4 metre lengths.
How many pieces will he get?

4 Get Real!

On her birthday, Bridget is given a big box of small sweets
called Little Diamonds.
She wants to find out how many sweets are in the box, but it
would take too long to count them.
A label on the box tells her that the total weight is 500 g.
She weighs 10 sweets. The weight of the 10 sweets is 0.4 g.

a How much does one sweet weigh?

b How many sweets are there in the box?

5 Hazel says that 48 ÷ 2 = 24, so 48 ÷ 0.2 = 2.4
Darren says 48 ÷ 2 = 24, so 4.8 ÷ 0.2 = 2.4
Harry says 48 ÷ 2 = 24, so 4.8 ÷ 2 = 2.4
Who is right? Give a reason for your answer.

6 €1 is worth £0.60. How many euro would you get for £7.50?

7 Here are two multiplagons.
On each straight line, the numbers in the circles multiply together
to make the number in the rectangle.
Your job is to copy and complete them by filling in the missing numbers.

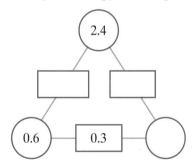

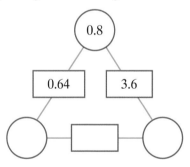

8 Carrie knows that 3.4 ÷ 0.4 = 8.5
Use this to copy and fill in the gaps in these questions.

a $34 ÷ 0.4 = \boxed{}$ **c** $340 ÷ \boxed{} = 8.5$ **e** $\boxed{} ÷ 0.04 = 8.5$

b $3.4 ÷ 4 = \boxed{}$ **d** $\boxed{} ÷ 4 = 0.85$ **f** $0.34 ÷ 8.5 = \boxed{}$

9 Start with a number less than 1.   *Example:*   0.8
Take it away from 1.   1 − 0.8   = 0.2
Divide the first number by the second.   0.8 ÷ 0.2 = 8 ÷ 2 = 4
Divide this number by one more than itself. 4 ÷ 5   = 0.8
You end up with the starting number!

Try this yourself, starting with

a 0.5   **b** 0.9   **c** 0.75

Check it with any number you like – although you will probably need a
calculator for more difficult examples.

Explore

◎ Draw a grid like this one:

$$\boxed{} . \boxed{} \div \boxed{0} . \boxed{}$$

◎ Roll a die

◎ Write the score in one of the empty boxes in your grid

◎ Roll the die twice more, writing the score in a box after each roll

◎ Work out the answer to the division

◎ Now try again – your aim is to get the highest possible answer

Investigate further

Learn 5 Fractions and decimals

Examples:

a Write 0.72 as a fraction.

To change a decimal to a fraction, just remember the place values.

Remember to use the place value of the *last* digit as the denominator

$$\begin{array}{ccc} \text{Units} & \text{Tenths} & \text{Hundredths} \\ 0 & . 7 & 2 \end{array} = \frac{72}{100} = \frac{18}{25}$$

The numerator and denominator have been divided by 4

$$0.72 = \frac{18}{25}$$

b Write $\frac{7}{8}$ as a decimal.

$\frac{7}{8}$ means $7 \div 8$.

$$\begin{array}{r} 0.875 \\ 8\overline{)7.^70^60^40} \end{array}$$

You can check your answers with a calculator

$$\frac{7}{8} = 0.875$$

Apply 5

1 Write these decimals as fractions, giving your answers in their simplest form.

a 0.6	**e** 0.65	**i** 0.375
b 0.32	**f** 0.125	**j** 0.015
c 0.9	**g** 0.55	
d 0.8	**h** 0.24	

2 Change these fractions to decimals.

   a $\frac{4}{5}$ **b** $\frac{7}{10}$ **c** $\frac{11}{20}$ **d** $\frac{11}{5}$

3 a Write these fractions as decimals.

     i $\frac{2}{5}$ **ii** $\frac{3}{8}$ **iii** $\frac{9}{20}$

  b Use your answers to part **a** to write the fractions in order of size, starting with the smallest.

4 Which of these fractions is closest to 0.67?

   a $\frac{3}{4}$ **b** $\frac{5}{8}$ **c** $\frac{3}{5}$

 HINT Write the fractions as decimals.

5 a What is 2.65 as a fraction?

  b What is $3\frac{7}{20}$ as a decimal?

6 Josh divided one number by another, and 2.375 was the answer. Both numbers were less than 20. What were the two numbers?

7 Write down five fractions that are equal to 0.4

8 Get Real!

At a school fête, some children decided to raise money with a 'Guess the weight of the cake' stall.
Amy guessed 3300 g, Tariq guessed 3.28 kg and Caroline guessed $3\frac{1}{5}$ kg.
The real weight was 3.237 kg. Who won?

9 Hilary says that $\frac{1}{8} = 1.8$

Nick says $\frac{3}{8} = 0.38$

Eleanor says $\frac{1}{10} = 0.10$

Jeff says $\frac{1}{20} = 0.20$

Who is correct? Correct the errors of the others.

10 Dan knows that $\frac{1}{8} = 0.125$
Use this answer to change these fractions to decimals:

   a $\frac{3}{8}$ **b** $1\frac{1}{8}$ **c** $\frac{5}{8}$ **d** $\frac{1}{16}$

11 Write these fractions as decimals. Be careful – they never end!
They are called recurring decimals.
Stop when you have reached a repeating digit or pattern of digits.

   a $\frac{2}{3}$ **b** $\frac{4}{11}$ **c** $\frac{3}{7}$

12 Find three fractions that fit all these rules:

  a All three fractions must have different denominators.

  b Each denominator must be less than 10.

  c The fraction must be greater than 0.2

  d The fraction must be less than 0.3

Explore

 Change all the unit fractions $(\frac{1}{2}, \frac{1}{3}, \frac{1}{4}, \frac{1}{5}, ...)$ up to $\frac{1}{10}$ to decimals

 Which give recurring decimals and which give terminating decimals?

> **Investigate further**

Decimals

The following exercise tests your understanding of this chapter, with the questions appearing in order of increasing difficulty.

1 Write the numbers below as decimal numbers.

a six tenths   i one half

b five hundredths   j one quarter

c eight thousandths   k two fifths

d twelve hundredths   l three per cent

e thirty-five thousandths

f five and nine tenths

g eleven and seven hundredths

h six and one tenth and two thousandths

2 What is the value of the 5 in the following numbers?

a 325   b 3526   c 42.57   d 3.00543   e 0.00005

3 Find the 'odd one out' of the following calculations.

a	b	c
$3.26 + 4.76 + 2.23$	$7.63 + 6.87 - 5.55$	$14.22 - 18.46 + 12.67$
$2.76 + 5.13 + 2.46$	$4.68 + 7.59 - 3.31$	$12.64 - 14.86 + 9.65$
$4.61 + 3.79 + 1.85$	$5.23 + 6.41 - 2.68$	$15.72 + 10.54 - 18.83$

4 a Jack and Jill fell down the hill. Jack fell 8.25 m and Jill fell 3.075 m further than Jack.
How far did Jill fall altogether?

b Captain Birdsi took his yacht on a 15.5 km race. The first leg of the race was 3.876 km and the second leg was 6.407 km. How long was the final leg of the race?

c Millie's puppy, Thomas, takes her for walks. On Sunday they walked 5.07 km; on Monday they walked 3.34 km; on Tuesday they walked 4.02 km; on Wednesday they walked 2.29 km; on Thursday they walked 4.8 km; on Friday they walked 2.67 km.
If Millie and Thomas walked a total of 26 km during the week, how far did they walk on Saturday?

d Tony wants to make the picture frame shown below.

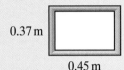

0.37 m

0.45 m

He has a suitable piece of wood one and a half metres long. Is there enough wood to make the frame? Depending on your answer, how much is either left over or needed?

e Delia is making cakes. She uses 0.135 kg sultanas in the first cake, 0.27 kg in the second, 0.185 kg in the third and 0.125 kg in the fourth. How many kilograms of sultanas does she have left from a 1 kg bag?

5 Work out the following:

a 0.4×0.7 **c** 0.02×0.235 **e** 4.025×12.5

b 2.1×11 **d** 20.4×4.3

f Mr Burton, the tailor, is cutting cloth for suits. Each suit takes 4.6 m of cloth. How much cloth is needed for 14 suits?

g One kilogram of nectarines costs £2.95. How much do 15 kilograms cost?

h Jane weighed 2.9 kg when she was born. On her first birthday she was 5.2 times as heavy. How heavy was she on her first birthday? Give your answer to the nearest 100 grams.

i A supermarket stocks boxes of the new breakfast cereal 'Chocobix'. Each packet of 'Chocobix' holds 625 g inside a cardboard box weighing 63 g. The supermarket shelf holds 36 of these packets. What is the total mass, in kilograms, on the shelf?

6 a Convert the following fractions to decimals:

i $\frac{3}{4}$ **ii** $\frac{1}{8}$ **iii** $\frac{4}{5}$ **iv** $\frac{7}{16}$ **v** $\frac{14}{25}$

b Convert the following decimals to fractions:

i 0.25   **ii** 0.375   **iii** 0.45   **iv** 0.16   **v** 0.6875

7 a Work out the following:

i $1.68 \div 0.4$ **iii** $16.9 \div 1.3$ **v** $49.2 \div 1.2$

ii $220 \div 0.05$ **iv** $6.25 \div 0.25$

b Road Runner travels 3.64 m in 0.7 seconds. How fast is this in metres per second?

c Naomi sees 54 suspect cells under her microscope in an area 0.06 cm². How many cells would she expect to find in an area of 1 cm²?

d A bag of sweets weighing 95 g includes wrappings of 0.5 g. Each sweet weighs 4.5 g. How many sweets are in the bag?

8 Perimeter and area

OBJECTIVES

G **Examiners would normally expect students who get a G grade to be able to:**

Find the perimeter of a shape by counting sides of squares

Find the area of a shape by counting squares

Estimate the area of an irregular shape by counting squares and part squares

Name the parts of a circle

F **Examiners would normally expect students who get an F grade also to be able to:**

Work out the area and perimeter of a simple rectangle, such as 3 m by 8 m

E **Examiners would normally expect students who get an E grade also to be able to:**

Work out the area and perimeter of a harder rectangle, such as 3.6 m by 7.2 m

D **Examiners would normally expect students who get a D grade also to be able to:**

Find the area of a triangle, parallelogram, kite and trapezium

Find the area and perimeter of compound shapes

Calculate the circumference of a circle to an appropriate degree of accuracy

Calculate the area of a circle to an appropriate degree of accuracy

C **Examiners would normally expect students who get a C grade also to be able to:**

Find the perimeter of a semicircle

Find the area of a semicircle

What you should already know ...

■ Multiply and divide one- and two-digit numbers

■ Find one half of a number

VOCABULARY

Shape – an enclosed space

Polygon – a closed two-dimensional shape made from straight lines

Triangle – a polygon with three sides

Quadrilateral – a polygon with four sides

Square – a quadrilateral with four equal sides and four right angles

Rectangle – a quadrilateral with four right angles, and the opposite sides equal in length

Area – the amount of enclosed space inside a shape

Perimeter – the distance around an enclosed shape

Dimension – the measurement between two points on the edge of a shape

Rhombus – a quadrilateral with four equal sides and opposite sides parallel

Parallelogram – a quadrilateral with opposite sides equal and parallel

Trapezium (pl. **trapezia**) – a quadrilateral with one pair of parallel sides

Kite – a quadrilateral with two pairs of equal adjacent sides

Pentagon – a polygon with five sides

Hexagon – a polygon with six sides

Octagon – a polygon with eight sides

Circle – a shape formed by a set of points that are always the same distance from a fixed point (the centre of the circle)

Diameter – a chord passing through the centre of a circle; the diameter is twice the length of the radius

Radius – the distance from the centre of a circle to any point on the circumference

Circumference – the perimeter of a circle

Chord – a straight line joining two points on the circumference of a circle

Semicircle – one half of a circle

Quadrant (of a circle) – one quarter of a circle

Learn 1   Perimeters and areas of rectangles

Examples: Find the perimeter and area of each of the following shapes.

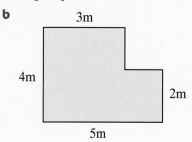

a

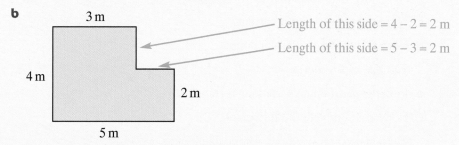

a

Perimeter = 5 + 3 + 5 + 3 = 16 cm ← Perimeter is the distance around the outside of a shape

Area (by counting the squares) = 15 cm² ← The units of area are squared

A quicker method than counting all the squares is to say that each row has 5 squares and there are 3 rows, so 5 × 3 = 15.

Area of a rectangle = length × width

b

3 m — Length of this side = 4 − 2 = 2 m

— Length of this side = 5 − 3 = 2 m

4 m

2 m

5 m

Perimeter = 5 + 2 + 2 + 2 + 3 + 4 = 18 cm

To find the area, divide the shape into rectangles:

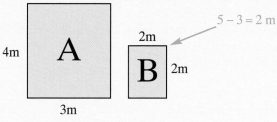

5 − 3 = 2 m

Area of A = 4 × 3 = 12 m²

Area of B = 2 × 2 = 4 m² ← The units of area are squared

Total area = 12 + 4 = 16 m²

Apply 1

1 a Without working out the areas look at these four rectangles and put
them in increasing order of size.

A

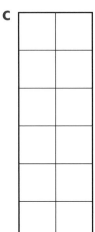

C

B

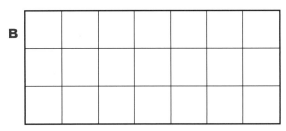

D

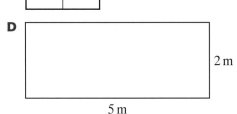

2 m

5 m

b Now work out the area of each rectangle.

c Is your answer to part **a** correct?

d Copy and complete the statement:

The area of a rectangle = ... × ...

e Find the perimeters of the four rectangles.

f Copy and complete the statement:

The perimeter of a rectangle can be found by ...

2 Apalia thinks that the area of a rectangle with length 4 cm and width 7 cm
is 28 cm. Is she correct? Give reasons for your answer.

3 How many different rectangles can you make using 100 one-centimetre squares?
Write down the dimensions of your rectangles.
(All dimensions must be whole numbers.)

4 Get Real!

Each student has a locker to put books in.

a What is the area of the base of the locker?

Christian turns each locker onto its side.

b What is the new area of the base of the locker?

c Which layout gives students more floor
space to put things in their lockers?

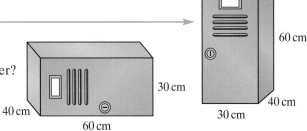

60 cm

30 cm

40 cm

40 cm

30 cm

60 cm

5 Find the perimeter of each of the following shapes:

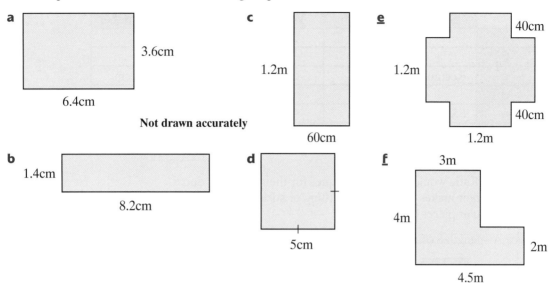

a 3.6cm 6.4cm

Not drawn accurately

b 1.4cm 8.2cm

c 1.2m 60cm

d 5cm

e 40cm 1.2m 40cm 1.2m

f 3m 4m 2m 4.5m

6 Find the area of each of the following shapes:

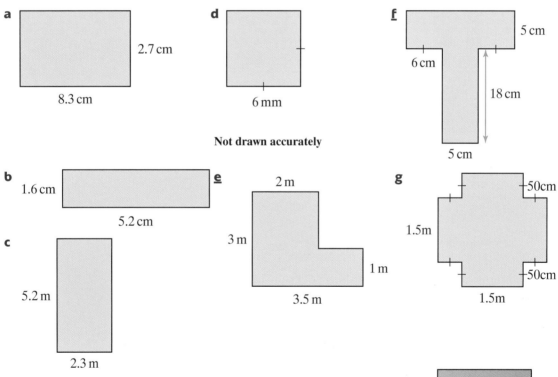

a 2.7 cm 8.3 cm

b 1.6 cm 5.2 cm

c 5.2 m 2.3 m

d 6 mm

Not drawn accurately

e 2 m 3 m 1 m 3.5 m

f 5 cm 6 cm 18 cm 5 cm

g 50cm 1.5m 50cm 1.5m

▦ 7 Get Real!

Anna is varnishing her door.
A small tin of varnish covers 1.15 m² and costs £1.20
A medium tin of varnish covers 3.4 m² and costs £3.20
Which tins of varnish should Anna buy for the cheaper option?

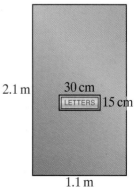

2.1 m 30 cm LETTERS 15 cm 1.1 m

8 Copy the following table and fill in the gaps.

	Shape	Length	Width	Area	Perimeter
a		4 cm	2 cm	8 cm²	
b		5 cm		15 cm²	
c	Rectangle	10 cm			26 cm
d		4 cm		16 cm²	
e	Square				20 cm
f		0.5 m	20 cm		

9 Get Real!

Andy and Katie would like name plaques for their bedroom doors.
The carpenter makes the letters by cutting or sticking together two types
of rectangular pieces of wood.

Piece A – length 8 cm, width 4 cm   Piece B – length 6 cm, width 4 cm

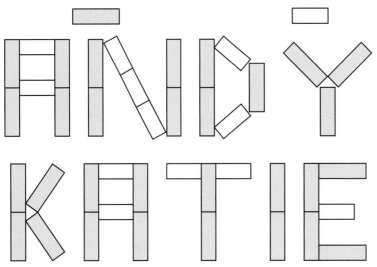

The carpenter charges 1p for every 2 cm² of wood.

a Which plaque is the cheapest to make?

b How much would your name plaque cost?

The carpenter can also put a finishing gold trim around the letters at a
charge of 1p per 2 cm.

c How much would your name plaque cost to trim?

Explore

Sally's hallway ceiling is made of 1 m square tiles. They have fallen down but some tiles have stayed together. The pieces lying on the floor are:

A B C D 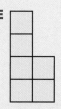 E

◎ Write down the area of each piece

◎ What is the total area of all five pieces?

◎ Make the pieces out of squared paper

◎ Try and fit them together to make the ceiling

◎ What are the dimensions of the rectangle?

◎ What is the area of this rectangle?

◎ Is it the same as you got for all five pieces?

HINT The ceiling is rectangular – use your last answer to help you!

Investigate further

Explore

◎ Betty's rabbit needs 3 m^2 of space

◎ Betty wants the space to be rectangular

◎ Betty also needs to buy wire fencing to keep foxes away from her rabbit

◎ Betty is running out of pocket money and wants to buy the smallest amount of wire fencing possible

◎ Betty thinks that the space will have to be 1 m × 3 m What is the perimeter?

◎ Betty's dad thinks a rectangle measuring 6 m × 0.5 m is better Does this have the same perimeter?

Investigate further

Learn 2   Perimeters and areas of triangles and parallelograms

Examples: Find the perimeter and area of the following shapes.

a Triangle

5 m   3 m   4.2 cm

Not drawn accurately

7 m

Perimeter = 7 + 5 + 4.2 = 16.2 m

Area = $\frac{1}{2} \times 7 \times 3 = 10.5$ m^2

Area of a triangle = $\frac{1}{2}$ × base × perpendicular height   Area = $\frac{1}{2} \times b \times h$

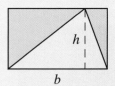

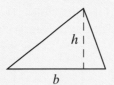

Two triangles can be joined together to make a rectangle with the same base and height. The area of a triangle is half the area of a rectangle with the same base and height.

b Parallelogram

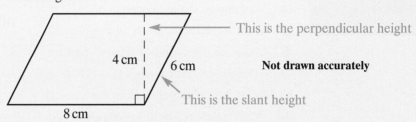

This is the perpendicular height

4 cm   6 cm   **Not drawn accurately**

This is the slant height

8 cm

Perimeter = 8 + 6 + 8 + 6 = 28 cm
Area = 8 × 4 = 32 cm^2

Area of a parallelogram = base × perpendicular height

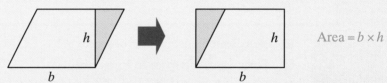

Area = $b \times h$

A parallelogram can be transformed into a rectangle as shown above. They both have the same area.

Apply 2

1 a Copy and complete the table for the following shapes made from 1 cm squares.

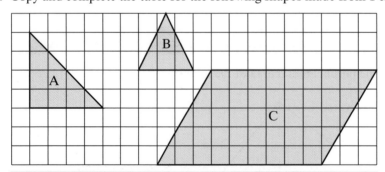

Shape	Name	Base (cm)	Perpendicular height (cm)	Area (cm^2)
A				
B				
C				

b Copy and complete the statements:

The area of a triangle = ... × ... × ...

The area of a parallelogram = ... × ...

2 Find the area of each of the following shapes:

a
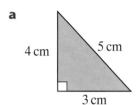

c
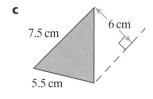

e
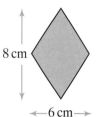

b
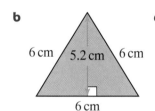

d
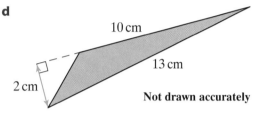

Not drawn accurately

f
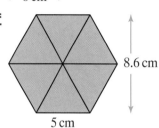

 3 Five students are trying to find the area of the following triangle:

- Sameera thinks the answer is 48 cm^2 because $6 \times 8 = 48$
- Bruce thinks the answer is 30 cm^2 because $\frac{1}{2} \times 6 \times 10 = 30$
- Cassie thinks the answer is 24 cm^2 because $6 + 8 + 10 = 24$
- Des thinks the answer is because 40 cm^2 because $\frac{1}{2} \times 8 \times 10 = 40$
- Elliot thinks the answer is 24 cm^2 because $\frac{1}{2} \times 6 \times 8 = 24$

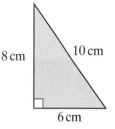

Who is correct? What mistakes have the other students made?

4 **a** The area of a parallelogram is 60 cm^2. Sketch five parallelograms with that area, showing the dimensions in each case.

  b The area of a triangle is 30 cm^2. Sketch five triangles with that area, showing the dimensions in each case.

5 Find the area of each of the following parallelograms:

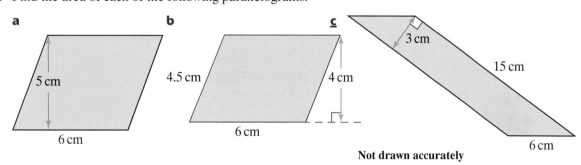

Not drawn accurately

6 Copy and fill in the gaps in the table.

	Shape (Parallelogram/Triangle)	Base	Perpendicular height	Area
a	Parallelogram	5 cm	4 cm	
b	Triangle	5 cm	4 cm	
c		10 cm	2 cm	10 cm²
d	Parallelogram	4 cm		8 cm²
e	Triangle	4 cm		8 cm²
f	Parallelogram	0.5 m	20 cm	

7 Get Real!

Reece wants to make a kite. The yellow silk costs £5 per square metre and the green silk costs £7 per square metre. Will he be able to buy enough silk with £3?

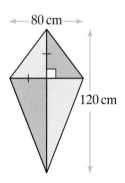
80 cm
120 cm

Explore

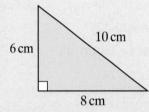
6 cm · 10 cm · 8 cm

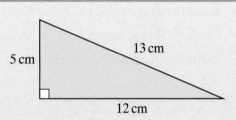
5 cm · 13 cm · 12 cm

- ◎ Calculate the perimeters of the two triangles
- ◎ Calculate the areas of the two triangles
- ◎ What do you notice?

Investigate further

Explore

- ◎ Calculate the area of the parallelogram
- ◎ Find the dimensions of parallelograms with an area that is one half of the area of the parallelogram shown
- ◎ Find the dimensions of parallelograms with an area that is one quarter of the area of the parallelogram shown

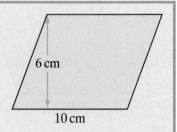
6 cm
10 cm

Investigate further

Learn 3   Areas of compound shapes

Example: Find the area of the following shape:

a Parallelogram

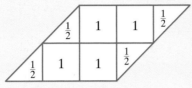

Area $= \frac{1}{2} + 1 + 1 + \frac{1}{2} + \frac{1}{2} + 1 + 1 + \frac{1}{2}$
$= 6 \text{ cm}^2$

Remember that the area of a parallelogram = base × perpendicular height $= 3 \times 2 = 6 \text{ cm}^2$

b Kite

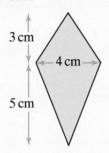

Area =

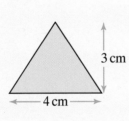   +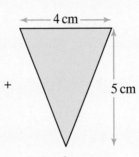

Area $= \frac{1}{2} \times 4 \times 3$
$= 6 \text{ cm}^2$

Area $= \frac{1}{2} \times 4 \times 5$
$= 10 \text{ cm}^2$

Total area $= 6 + 10 = 16 \text{ cm}^2$

In general the area of a kite $= \frac{1}{2}$ height × width
$= \frac{1}{2} \times (3 + 5) \times 4$
$= \frac{1}{2} \times 8 \times 4$
$= 16 \text{ cm}^2$

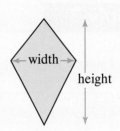

c Trapezium

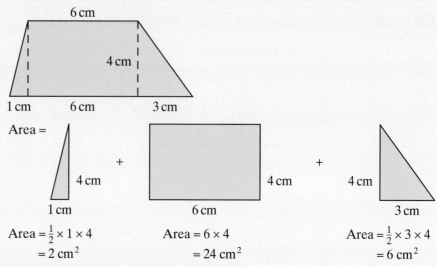

Area =

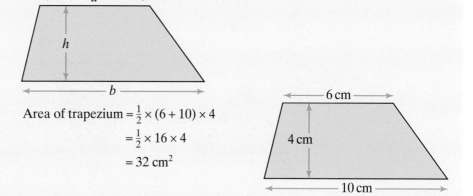

$Area = \frac{1}{2} \times 1 \times 4$
$= 2 \text{ cm}^2$

$Area = 6 \times 4$
$= 24 \text{ cm}^2$

$Area = \frac{1}{2} \times 3 \times 4$
$= 6 \text{ cm}^2$

Total area = 2 + 24 + 6 = 32 cm²

In general the area of a trapezium = $\frac{1}{2} \times$ (sum of parallel sides) $\times h$

Area of trapezium = $\frac{1}{2} \times (6 + 10) \times 4$
$= \frac{1}{2} \times 16 \times 4$
$= 32 \text{ cm}^2$

Apply 3

1 Estimate the area of the island where each square represents one square mile.

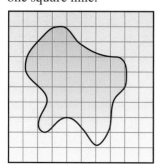

2

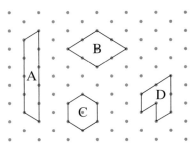

    a Which of the shapes have the same area?

    b Which of the shapes has the largest area?

3   **a** Find the area of each of the following shapes.

    b Find the perimeters of each of the following shapes.

i

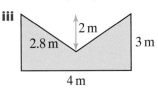

5.4 cm

2 cm

13 cm

10 cm

5 cm

iii

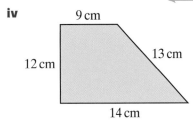

2 m

2.8 m

3 m

4 m

Not drawn accurately

v

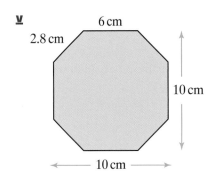

2.8 cm

6 cm

10 cm

10 cm

ii

10 cm

5.7 cm

4 cm

8 cm

14 cm

iv

9 cm

12 cm

13 cm

14 cm

4 Roberta is finding the area of a hexagon. She spots that she can split it
into a rectangle and two triangles:

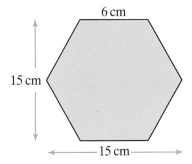

6 cm

15 cm

15 cm

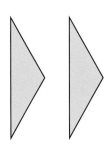

Area of rectangle = $6 \times 15 = 90$ cm^2
Area of triangle = $15 \times 4.5 = 67.5$ cm^2

Area of hexagon = $90 + 2 \times 67.5 = 225$ cm^2

Do you agree with Roberta? Give reasons for your answer.

5 Get Real!

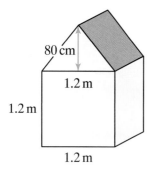

George is varnishing the front of Fred's doghouse. How many tins of varnish does he need? (The label on the tin has instructions that 1 litre of paint covers 0.5 m².)

6 Get Real!

Jane has made an 'EXIT' sign using a large piece of cardboard and two types of small rectangles to make the letters. Calculate the area of grey card she needs to paint.

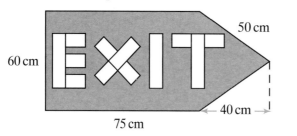

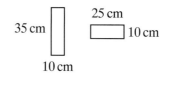

7 Find the area of each of the following shapes:

a

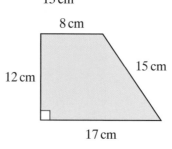

c

e

b

d

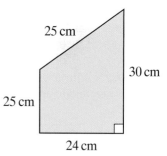

8 Sketch five trapezia with the area 25 cm², stating clearly the dimensions in each case.

Explore

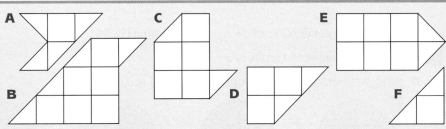

- ◎ Write down the area of each piece
- ◎ What is the total area of all six pieces?
- ◎ Make the pieces out of squared paper
- ◎ Try and fit them together to make a parallelogram
- ◎ What are the dimensions of the parallelogram?
- ◎ What is the area of this parallelogram?
- ◎ Is it the same as you got for all six pieces?

Investigate further

Explore

- ◎ Make two copies of the trapezium shown
- ◎ Try and fit the two pieces together to make a rectangle or parallelogram
- ◎ What is the area of the shape you have made?
- ◎ Deduce the area of the trapezium

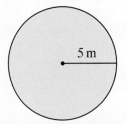

Investigate further

Learn 4   Circumferences of circles

Examples: Calculate the circumference of this circle:

a leaving your answer in terms of π

b giving your answer to 3 significant figures.

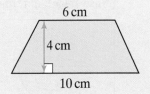

Use the fact that $c = \pi \times d$ where c is the circumference and d is the diameter

Diameter = 2 × radius

a Circumference = $\pi d = \pi \times 10 = 10\pi$ m

You may be asked to leave your answers in terms of π on the non-calculator paper

b Circumference = $\pi d = \pi \times 10 = 31.4$ m (3 s.f.)

The calculator gives more decimal places but you need to round to an appropriate degree of accuracy (usually 3 s.f. or 2 d.p.)

95

Apply 4

1 Calculate the circumference of each of the following circles:

 i leaving your answer in terms of π

ii giving your answer to an appropriate degree of accuracy.

a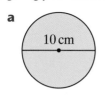
10 cm

b
4 cm

c
4 m

d
5 mm

2 Get Real!

The London Eye has a diameter of 135 m and takes approximately half an hour to make one complete revolution. How far has the base of a capsule travelled in:

a 30 minutes

b 15 minutes

c 1 hour?

3 The circumference of a circle is π times d.
Can you find a correct line of three?
(Answers are in terms of π or to 1 decimal place.)

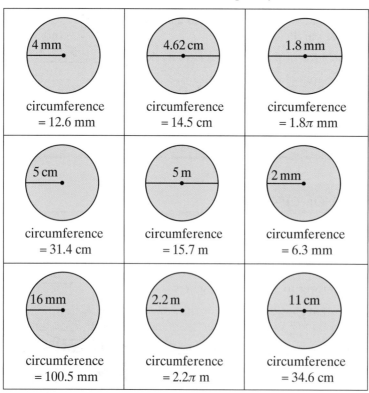

4 mm circumference = 12.6 mm	4.62 cm circumference = 14.5 cm	1.8 mm circumference = 1.8π mm
5 cm circumference = 31.4 cm	5 m circumference = 15.7 m	2 mm circumference = 6.3 mm
16 mm circumference = 100.5 mm	2.2 m circumference = 2.2π m	11 cm circumference = 34.6 cm

4 Calculate the total perimeter of the following shapes:

 i leaving your answers in terms of π

ii giving your answers to an appropriate degree of accuracy.

a

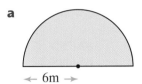

← 6m →

b

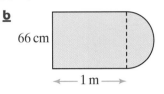

66 cm

←— 1 m —→

c

6 cm

d

45°

10 cm

5 Get Real!

Ahmed lives 1500 m from school. The diameter of his bike's wheel is 80 cm.

a Find the circumference of the wheel.

b How many complete revolutions does the wheel turn during Ahmed's journey to school?

6 Get Real!

Jack and Susan have a race around the track shown below.
Jack is in lane 1 and Susan in lane 8.

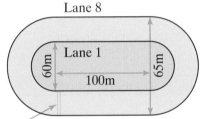

Lane 8

Lane 1

60m

100m

65m

START

a How far does Jack run?

b How far does Susan run?

c How can you make the race fair?

7 Copy and complete the following table, giving your answers to an appropriate degree of accuracy.

Radius	Diameter	Circumference
	5 cm	
4 m		
		10 mm
		15π cm

8 Get Real!

Ahmed's CDs have a circumference of 40 cm.
He wants to make square covers for them.
Find, correct to 1 d.p., the dimensions of the smallest square into which the CDs will fit.

<u>9</u> Get Real!

Radius is a prime number

Radius is a multiple of 5

Radius is a factor of 36

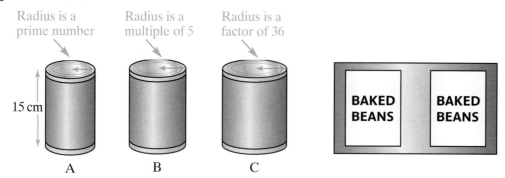

15 cm

A   B   C

BAKED BEANS

BAKED BEANS

The labels have fallen off three tins.
The area of the 'Baked Beans' label is 942.5 cm^2 (1 d.p.).

a Which tin does it fit – A, B or C?

b Is the area of the label approximately 1 square metre?
Give a reason for your answer.

Explore

◎ Find five circular objects

◎ For each object, measure the circumference and the diameter

◎ Copy and complete the table

Object	Circumference (c cm)	Diameter (d cm)	c ÷ d

Investigate further

Learn 5   Areas of circles

Examples: Calculate the area of this circle:

a leaving your answer in terms of π

Use the fact that A $= \pi \times r^2$ where A is the area and r is the radius

b giving your answer to 3 significant figures.

5 cm

You may be asked to leave your answers in terms of π on the non-calculator paper

a Area $= \pi r^2 = \pi \times 5^2 = 25\pi$ cm^2

The units are squared for area

b Area $= \pi r^2 = \pi \times 5^2 = \pi \times 25 = 78.5$ cm^2 (3 s.f.)

The calculator gives more decimal places but you need to round to an appropriate degree of accuracy (usually 3 s.f. or 2 d.p.)

Apply 5

1 Calculate the area of each of the following circles:

 i leaving your answer in terms of π

    ii giving your answer to 2 d.p.

a
10 cm

b
4 cm

c
4 m

d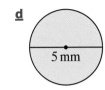
5 mm

2 Joy and Jan are finding the area of a CD.

    a This is Joy's method:

      Area $= \pi \times d = \pi \times 10 = 31.42 \text{ cm}^2$ (to 2 d.p.)

      Do you agree with Joy's answer? Justify your answer.

    b This is Jan's method:

      Area $= \pi r^2 = \pi \times 5^2 = 246.74 \text{ cm}^2$ (to 2 d.p.)

      Calculator buttons used: $\boxed{\pi}$ $\boxed{\times}$ $\boxed{5}$ $\boxed{=}$ $\boxed{x^2}$ $\boxed{=}$

      Do you agree with Jan's answer? Justify your answer.

10 cm

3 Copy this and match each circle with the correct area.
    Fill in the missing areas and the missing radius.

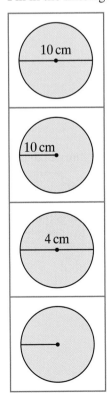
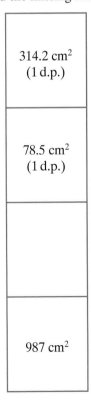

10 cm	314.2 cm² (1 d.p.)
10 cm	78.5 cm² (1 d.p.)
4 cm	
	987 cm²

4 Calculate the total area of each of the following shapes:

 **a** leaving your answer in terms of π

b giving your answer to an appropriate degree of accuracy.

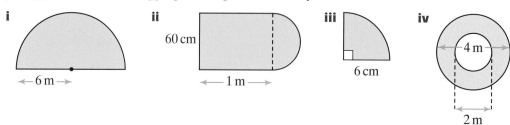

i   6 m

ii   60 cm   1 m

iii   6 cm

iv   4 m   2 m

5 Complete the following table, giving your answers to an appropriate degree of accuracy.

Radius	Diameter	Area
	5 cm	
4 m		
		10 mm²
		16π cm²

6 Get Real!

Steve wants to paint his favourite computer game on his bedroom wall. His parents aren't so keen! They will only allow him to do it if the black paint covers no more than two thirds of the wall.

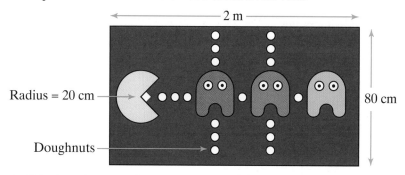

2 m

Radius = 20 cm

80 cm

Doughnuts

a Calculate the area of one monster (including the eyes).

b Calculate the total area of the doughnuts if each one has a radius of 5 cm.

c Calculate the area covered by the yellow doughnut eater.

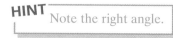

HINT Note the right angle.

d Calculate the total area covered by the monsters, doughnuts and doughnut eater.

e Calculate the area which will be black?

f Will Steve's parents allow him to paint his wall?

Monster details

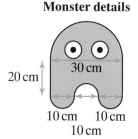

30 cm

20 cm

10 cm   10 cm

10 cm

Explore

- ◎ Divide a circle into six equal pieces and cut out the pieces
- ◎ Try to make the following shape:

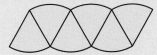

- ◎ Measure the length and height of your shape
- ◎ Work out the approximate area of the shape
- ◎ Repeat with the same size circle but this time divided into 10 pieces

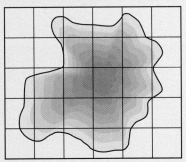

Investigate further

Perimeter and area

ASSESS

The following exercise tests your understanding of this chapter, with the questions appearing in order of increasing difficulty.

1 a This coffee stain is divided up into centimetre squares.
Estimate its area.

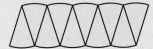

b The diagram shows a map of Jersey.
Each full square represents 1 square mile.
Estimate the area of Jersey to the nearest 5 square miles.

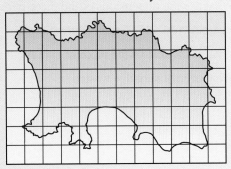

2 Accurately draw the diagram below on a piece of card or paper.

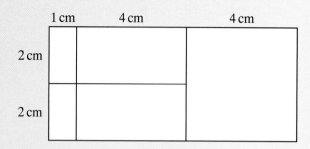

a Write down the area and perimeter of the large rectangle.

Cut the shape into the five pieces shown.
Rearrange the five pieces to make a square.

b Write down the area and perimeter of the square.

c What do you notice?

3

The diagram shows Tom and Ali's garden. The barbecue area is a square
and the path is 1 m wide. Find the area and perimeter of:

a the patio   d the path

b the lawn   e the flower bed

c the barbecue   f the whole garden.

4 A rectangular room measuring 3 m 81 cm by 4 m 65 cm is to be carpeted.
The carpet is cut from a roll 4 m wide and the length can only be cut
to an exact number of metres.

a What are the dimensions of the piece of carpet cut from the roll?

The carpet costs £17.95 per square metre.

b What is the cost of the piece of carpet cut from the roll?

c What area of carpet is wasted?

5 Find the areas of the following shapes.

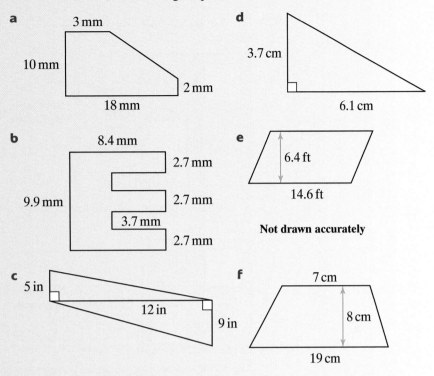

a
3 mm
10 mm
2 mm
18 mm

b
8.4 mm
2.7 mm
9.9 mm
2.7 mm
3.7 mm
2.7 mm

c
5 in
12 in
9 in

d
3.7 cm
6.1 cm

e
6.4 ft
14.6 ft

Not drawn accurately

f
7 cm
8 cm
19 cm

6 a Find the circumference of a circle of diameter 12.6 cm. (Take $\pi = 3.14$)

   b A garden reel contains 30 m of hosepipe. The reel has a diameter of 20 cm. Calculate the number of times the reel rotates when the complete length of the hosepipe is unwound.
(You may ignore the thickness of the hosepipe.)

7 a A washer has an outer radius of 3.6 cm and a hole of radius 0.4 cm.

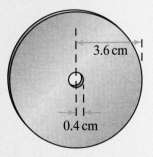

3.6 cm

0.4 cm

Calculate the area of the face of the washer, giving your answer:

i in terms of π

ii to 1 decimal place.

   b Three thin silver discs, of radii 4, 7 and 10 cm, are melted down and recast into another disc of the same thickness. Find the radius of this disc.

Try a real past exam question to test your knowledge:

8 a Two squares of side 4 cm are removed from a square of side 12 cm as shown.

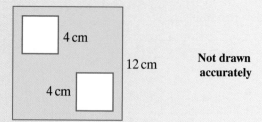

Work out the shaded area.

b Two squares of side x cm are removed from a square of side $3x$ cm as shown.

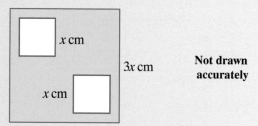

Work out the fraction of the large square which remains.
Give your answer in its simplest form.

Spec B Modular, Module 5, Nov 04

9 Fractions

OBJECTIVES

G **Examiners would normally expect students who get a G grade to be able to:**

Find equivalent fractions

F **Examiners would normally expect students who get an F grade also to be able to:**

Simplify fractions, such as $\frac{24}{36}$

Arrange fractions in order of size

E **Examiners would normally expect students who get an E grade also to be able to:**

Work out fractions of quantities, such as $\frac{5}{8}$ of £20

Find one number as a fraction of another

Do calculations with simple fractions involving addition and multiplication

Convert fractions such as $\frac{3}{8}$ to decimals

D **Examiners would normally expect students who get a D grade also to be able to:**

Do calculations with simple fractions involving subtraction

C **Examiners would normally expect students who get a C grade also to be able to:**

Do calculations with simple fractions involving division

Do calculations with mixed numbers

What you should already know ...

■ Understand basic fractions

■ Understand numerator and denominator

Fraction or **simple fraction** or **common fraction** or **vulgar fraction** – a number written as one whole number over another, for example, $\frac{3}{8}$ (three eighths), which has the same value as $3 \div 8$

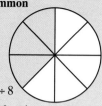

Numerator – the top number in a fraction

Numerator —————→ $\frac{3}{8}$ ←————— Denominator

Denominator – the bottom number in a fraction

Unit fraction – a fraction with a numerator of 1, for example, $\frac{1}{5}$

Proper fraction – a fraction in which the numerator is smaller than the denominator, for example, $\frac{5}{13}$

Improper fraction or **top-heavy fraction** – a fraction in which the numerator is bigger than the denominator, for example, $\frac{13}{5}$, which is equal to the mixed number $2\frac{3}{5}$

Mixed number or **mixed fraction** – a number made up of a whole number and a fraction, for example, $2\frac{3}{5}$, which is equal to the improper fraction $\frac{13}{5}$

Decimal fraction – a fraction consisting of tenths, hundredths, thousandths, and so on, expressed in a decimal form, for example, 0.65 (6 tenths and 5 hundredths)

Equivalent fraction – a fraction that has the same value as another, for example, $\frac{3}{5}$ is equivalent to $\frac{30}{50}, \frac{6}{10}, \frac{60}{100}, \frac{15}{25}, \frac{1.5}{2.5}, \cdots$

Simplify a fraction or **express a fraction in its simplest form** – to change a fraction to the simplest equivalent fraction; to do this divide the numerator and the denominator by a common factor (this process is called cancelling or reducing or simplifying the fraction)

Learn 1 Equivalent fractions

Example:

Write down a fraction equivalent to $\frac{9}{12}$

This diagram shows a number line from 0 to 1 split up into twelfths and into quarters.

| $\frac{1}{12}$ | $\frac{2}{12}$ | $\frac{3}{12}$ | $\frac{4}{12}$ | $\frac{5}{12}$ | $\frac{6}{12}$ | $\frac{7}{12}$ | $\frac{8}{12}$ | $\frac{9}{12}$ | $\frac{10}{12}$ | $\frac{11}{12}$ | $\frac{12}{12}$ |

The shading shows that nine twelfths is the same as three quarters, $\frac{9}{12} = \frac{3}{4}$

In other words, $\frac{9}{12}$ and $\frac{3}{4}$ are equivalent fractions.

The fractions can be changed into one another: $\frac{9}{12} \overset{\div 3}{\underset{\div 3}{=}} \frac{3}{4}$ and $\frac{3}{4} \overset{\times 3}{\underset{\times 3}{=}} \frac{9}{12}$

The value of a fraction does not change if you multiply the top number (numerator) and the bottom number (denominator) by the same number.

The value of a fraction does not change if you divide the numerator and the denominator by the same number.

Apply 1

You should be able to do all these questions without a calculator, but you may want to use a calculator to check your work and to speed things up.
Find out how to use the $\boxed{a^{b/c}}$ key if you have one on your calculator.

1 Copy and complete these equivalent fraction statements.

a $\dfrac{5}{6} = \dfrac{\square}{12}$

c $\dfrac{6}{12} = \dfrac{\square}{4} = \dfrac{1}{2}$

b $\dfrac{3}{12} = \dfrac{\square}{4}$

d $\dfrac{12}{12} = \dfrac{4}{\square} = 1$

2

$\frac{1}{10}$	$\frac{2}{10}$	$\frac{3}{10}$	$\frac{4}{10}$	$\frac{5}{10}$	$\frac{6}{10}$	$\frac{7}{10}$	$\frac{8}{10}$	$\frac{9}{10}$	$\frac{10}{10}$
$\frac{1}{5}$		$\frac{2}{5}$		$\frac{3}{5}$		$\frac{4}{5}$		$\frac{5}{5}$	
$\frac{1}{2}$					$\frac{2}{2}$				

Use this diagram to copy and complete these equivalent fraction statements.

a $\dfrac{3}{5} = \dfrac{6}{\square}$

b $\dfrac{\square}{\square} = \dfrac{1}{2}$

c $\dfrac{\square}{10} = \dfrac{\square}{5} = \dfrac{\square}{2} = 1$

d Use the diagram to write down another statement about equivalent fractions.

3 Use this circle diagram to write some equivalent fraction statements.

Each part is one twelfth

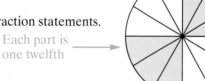

4 Copy and complete these equivalent fraction statements.
Use diagrams to help you if you need to.

a $\dfrac{3}{4} = \dfrac{\square}{8}$

d $\dfrac{8}{\square} = \dfrac{4}{\square} = \dfrac{2}{\square} = 1$

b $\dfrac{2}{\square} = \dfrac{1}{4}$

e $\dfrac{1}{3} = \dfrac{\square}{6} = \dfrac{\square}{9} = \dfrac{\square}{12} = \dfrac{\square}{15}$

c $\dfrac{1}{\square} = \dfrac{2}{\square} = \dfrac{4}{8}$

5 Jack says that $\frac{3}{5} = \frac{6}{8}$

Is he correct? Explain your answer.

6 Mark says 'If the numerator and the denominator of a fraction are the same, the fraction is equal to 1.'
Is Mark correct? Explain your answer.

7 Write down five fractions that are equivalent to $\frac{2}{3}$

8 Find the odd fraction out in this list: $\frac{8}{10}, \frac{6}{8}, \frac{4}{5}, \frac{16}{20}$

Now do the question by changing each of the fractions to decimal form.

9 Write the simplest possible equivalent fraction for each of these fractions.

(Make sure that you know how to do these without a calculator but also find out how to use a calculator to help you simplify fractions.
The simplest way is to use the $\boxed{a^{b/c}}$ key if you have one.)

a $\dfrac{8}{16}$ **f** $\dfrac{18}{36}$ **k** $\dfrac{37}{74}$ **p** $\dfrac{30}{300}$

b $\dfrac{20}{30}$ **g** $\dfrac{12}{16}$ **l** $\dfrac{72}{120}$ **q** $\dfrac{100}{75}$

c $\dfrac{15}{20}$ **h** $\dfrac{15}{24}$ **m** $\dfrac{15}{150}$

d $\dfrac{24}{36}$ **i** $\dfrac{30}{72}$ **n** $\dfrac{2.5}{3.5}$

e $\dfrac{75}{100}$ **j** $\dfrac{45}{100}$ **o** $\dfrac{\frac{1}{2}}{\frac{3}{2}}$

This is a 'top-heavy' (improper) fraction but it can still be simplified

10 Write down three equivalent fraction statements about fifths, tenths and twentieths.

11 Write down an equivalent fraction statement about hundredths, tenths and fifths.

12 Get Real!

A bar of chocolate is spilt into 24 equal pieces and Sam eats 4 of them.

Joe has another bar of chocolate the same size that is split into 30 equal pieces.

How many pieces of that bar should Joe eat so that he eats the same amount of chocolate as Sam?

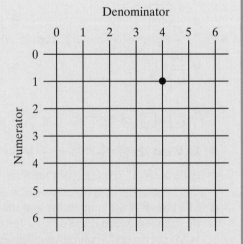

Explore

Fractions can be marked on a grid like this one. The dot shows $\frac{1}{4}$, because its numerator is 1 and its denominator is 4.

◎ Make a grid like this, going up to 12 in each direction

◎ Mark all the fractions that are equivalent to 1 ($\frac{1}{1}, \frac{2}{2}, \frac{3}{3}$, etc.) What do you notice?

◎ Mark two more sets of equivalent fractions
Compare one set with the other
What is the same and what is different?

◎ Where do top-heavy (improper) fractions appear on the grid?

◎ How can this grid help you to arrange fractions in order?

◎ How can the grid help you to simplify fractions?

Denominator

0 1 2 3 4 5 6

Numerator

0
1
2
3
4
5
6

Investigate further

Learn 2 Arranging fractions

Example: Arrange the following fractions in order of size, largest first.

$$\frac{2}{3}, \frac{1}{4}, \frac{5}{6}, \frac{1}{12}, \frac{1}{2}$$

In order to put fractions in order of size, first change them all to the same denominator.

All the denominators are factors of 12, so all the fractions can be converted to twelfths

Change the fractions to twelfths: $\frac{2}{3} \overset{\times 4}{\underset{\times 4}{=}} \frac{8}{12}$ and so on.

So the list $\frac{2}{3}, \frac{1}{4}, \frac{5}{6}, \frac{1}{12}, \frac{1}{2}$ becomes $\frac{8}{12}, \frac{3}{12}, \frac{10}{12}, \frac{1}{12}, \frac{6}{12}$

which in order of size is $\frac{1}{12}, \frac{3}{12}, \frac{6}{12}, \frac{8}{12}, \frac{10}{12}$

Apply 2

All these questions can be done without a calculator and you should make sure that you are able to do without. Also make sure that you can use your calculator to speed up and to check your work, using your $\boxed{a^{b/c}}$ *key if you have one.*

1 Change these fractions to fifteenths and then arrange them in order of size.

$$\frac{7}{15}, \frac{2}{5}, \frac{2}{3}, \frac{4}{15}$$

2 Change these fractions to hundredths and then arrange them in order of size.

$$\frac{7}{10}, \frac{2}{5}, \frac{27}{50}, \frac{16}{25}$$

Fractions with a denominator of 100 are percentages, for example, $\frac{70}{100}$ is 70%

3 Arrange these fractions in order of size: $\frac{2}{3}, \frac{2}{5}, \frac{1}{2}, \frac{7}{10}$

4 a What is a good common denominator for arranging halves, eighths and quarters in order?

 b Explain why it is not easy to change fifths to twelfths.

5 a Carlo says 'Four fifths is smaller than four sevenths.' Is he correct?

 b Jack says 'Eight ninths is smaller than nine tenths.' Is he correct?

6 a Find three fractions that are bigger than one quarter and smaller than one half.

 b Find a fraction between nine tenths and one.

Explore

◉ Choose any simple fraction, for example $\frac{3}{4}$

◉ Add one to the numerator and one to the denominator, for example $\frac{3+1}{4+1} = \frac{4}{5}$

◉ Find out whether the new fraction is less than, bigger than or the same size as the original fraction

Investigate further

Learn 3 Fractions of quantities

Example:

There are 54 people in a choir. Two thirds of the choir are women. How many women are there in the choir?

To find the number of women in the choir you need to find $\frac{2}{3}$ of 54.

To work out two thirds of 54, first work out one third of 54, then multiply the answer by two to find two thirds.

$\frac{1}{3}$ of $54 = 54 \div 3 = 18$ ◄—— $\frac{1}{3}$ of 54 ——►

So

18	36	54

$\frac{2}{3}$ of $54 = 18 \times 2 = 36$ ◄——————— $\frac{2}{3}$ of 54 ———————►

There are 36 women in the choir.

It doesn't matter whether you divide by three and then multiply by two or multiply by two and then divide by three – the result is the same.

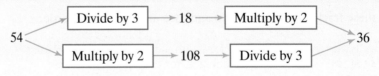

54 → Divide by 3 → 18 → Multiply by 2 → 36

54 → Multiply by 2 → 108 → Divide by 3 → 36

Apply 3

1 Work out:

a $\frac{2}{3}$ of 12

b $\frac{2}{3}$ of £15

c $\frac{2}{3}$ of £30

d $\frac{2}{3} \times £18$

e $150 \times \frac{2}{3}$

f $\frac{2}{3}$ of an hour (find the answer in minutes)

2 Explain how you can use the answer to question **1b** to find the answer to question **1c**.

3 a Two thirds of a number is 10. What is the number?

  b Two thirds of a number is 20. What is the number?

4 Find three quarters of:

   a 100

   b 50

   c an hour (answer in minutes)

   d a metre (answer in centimetres)

   e a kilogram (answer in grams)

5 Three quarters of a number is 15. What is the number?

6 Find $\frac{4}{3}$ of:

   a £15   **b** £30   **c** £18   **d** 150   **e** an hour

7 Which of these calculations will work out $\frac{3}{5}$ of 44?
Write down all of the calculations that apply.

   a $44 \div 5 \times 3$ **c** $44 \times 5 \div 3$ **e** $44 \div 100 \times 60$

   b $44 \div 3 \times 5$ **d** $44 \times 3 \div 5$ **f** $44 \times 10 \div 6$

8 Get Real!

A type of bronze used to make coins is made up of copper, tin and zinc.

$\frac{95}{100}$ of the bronze is copper, $\frac{4}{100}$ is tin and the rest is zinc.

How much tin is there in 2 kg of bronze?

Explore

  ◎ Find $\frac{3}{4}$ of 20

  ◎ Now find $\frac{4}{3}$ of the result

  ◎ What do you notice?

  ◎ Can you explain what has happened?

Investigate further

Learn 4   One quantity as a fraction of another

A total of 28 students took a maths test and 16 students passed the test.
What fraction of the students passed the test?

This is the number of students who took the test → $\dfrac{16}{28}$ ← This is the number of students who passed the test

The fraction $\frac{16}{28}$ can be simplified to $\frac{4}{7}$.   Reminder: $\dfrac{16}{28} \overset{\div 4}{\underset{\div 4}{=}} \dfrac{4}{7}$

So $\frac{4}{7}$ of the class passed the test and $\frac{3}{7}$ of the class did not pass it. ← These results could also be expressed as decimals or percentages

In everyday life, percentages are usually used

Apply 4

You should be able to do all these without a calculator.
You may like to use your calculator to check your answers.

1 If there are 28 students in the class, what fraction of them passed the test if the number passing was:

   a 14   **b** 12   **c** 13   **d** 18   **e** 20?

   f If the fraction passing the test was $\frac{3}{4}$, how many students passed?

   g Explain why you should never get an improper (top-heavy) fraction in a question like this.

2 Here is a list of the test marks of a class of 30 students, arranged in order.

     22 25
     30 33 37
     42 43 46 46
     53 54 55 55 56
     61 61 63 64 64 67 68 68 69
     73 75 78 79
     81 87
     95

   a What fraction of the students got under 40 marks?

   b What fraction of the students got a mark in the sixties?

   c If the pass mark was 50 marks what fraction of the students would pass the test?

   d What should the pass mark be for two thirds of the students to pass the test?

   e What mark separates the top tenth of the class from the rest?

3 Get Real!

A T-shirt costing £7.50 has its price reduced by £1.50 in a sale.

a What fraction reduction is this?

b What fraction of the original price is the sale price?

c If the price of the T-shirt was reduced by $\frac{1}{3}$ what would the sale price be?

d If the sale price of a T-shirt was £6 after a reduction of $\frac{1}{3}$, what was the original price?

4 This graph shows the price of petrol in different countries.
The wholesale price and the taxes make up the price at the pump.

a Estimate what fraction of its total petrol price each country pays in tax.

b Which country pays the highest fraction in tax?

c Which country pays the lowest fraction in tax?

d The prices are given in US dollars. Would it affect your answers if the prices were given in euro? Explain your answer.

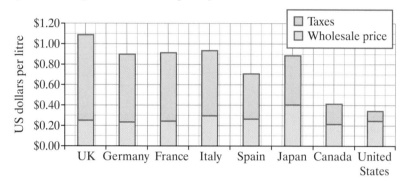

Explore

◎ In this diagram, what fraction of the outside square is black?

◎ What fraction is white?

◎ Now another black square has been added

◎ What fraction of the outside square is the new black square?

◎ What fraction of the outside square is black now?

◎ What fraction is white?

◎ Now another white square has been added

◎ What fraction of the outside square is the new white square?

◎ What fraction of the outside square is black now?

◎ What fraction is white?

Investigate further

Learn 5 Adding and subtracting fractions

Examples:

a Calculate $\frac{5}{6} + \frac{3}{4}$

Fractions cannot be added (or subtracted) unless they have the same denominator.

$\frac{5}{6} + \frac{3}{4}$ ←————— Change both the fractions to twelfths

$= \frac{10}{12} + \frac{9}{12}$ ←————— When the fractions have both been changed to twelfths, add to find the total number of twelfths

$= \frac{19}{12}$

$= 1\frac{7}{12}$ ←————— Simplify the answer by writing it as a mixed number

Reminder: $\frac{5}{6} \overset{\times 2}{=} \frac{10}{12}$ and $\frac{3}{4} \overset{\times 3}{=} \frac{9}{12}$

b Calculate $\frac{5}{6} - \frac{3}{4}$

$\frac{5}{6} - \frac{3}{4}$

$= \frac{10}{12} - \frac{9}{12}$

$= \frac{1}{12}$

c Calculate $1\frac{5}{6} + 2\frac{3}{4}$

$1\frac{5}{6} + 2\frac{3}{4}$

$= 1 + \frac{5}{6} + 2 + \frac{3}{4}$

$= 3 + \frac{10}{12} + \frac{9}{12}$

$= 3\frac{19}{12} = 4\frac{7}{12}$

d Calculate $2\frac{1}{6} - \frac{3}{4}$

$2\frac{1}{6} - \frac{3}{4}$

$= 2 + \frac{2}{12} - \frac{9}{12}$

$= 2 - \frac{7}{12}$ ←————— Change one of the units to twelfths to do this subtraction

$= 1\frac{5}{12}$ $2 - \frac{7}{12} = 1 + 1 - \frac{7}{12} = 1 + \frac{12}{12} - \frac{7}{12} = 1\frac{5}{12}$

Apply 5

Questions like this will be on the non-calculator paper so make sure you can do them without using your calculator.

1 Work out: **a** $\frac{3}{4} + \frac{2}{3}$ **b** $\frac{3}{4} - \frac{2}{3}$ **c** $\frac{5}{6} + \frac{2}{5}$ **d** $\frac{5}{6} - \frac{2}{5}$

2 Work out: **a** $3\frac{3}{4} + 1\frac{2}{3}$ **b** $3\frac{3}{4} - 1\frac{2}{3}$ **c** $2\frac{5}{6} + 1\frac{2}{5}$ **d** $2\frac{5}{6} - 1\frac{2}{5}$

3 Work out: **a** $3\frac{2}{3} + 1\frac{3}{4}$ **b** $3\frac{2}{3} - 1\frac{3}{4}$ **c** $2\frac{2}{5} + 1\frac{5}{6}$ **d** $2\frac{2}{5} - 1\frac{5}{6}$

4 Sue says 'I add up fractions like this: $\frac{5}{6} + \frac{2}{5} = \frac{5+2}{6+5} = \frac{7}{11}$,'
Is Sue right? Explain your answer.

5 Find two fractions with:

a a sum of $1\frac{1}{4}$

b a difference of $1\frac{1}{4}$

c a sum of $3\frac{1}{3}$

d a difference of $3\frac{1}{3}$

6 Get Real!

Anne is making custard, which needs $\frac{1}{3}$ of a cup of sugar.

Then she makes biscuits, which need $\frac{3}{4}$ of a cup of sugar.

Anne only has 1 cup of sugar.
Does she have enough to make the custard and the biscuits?
Show how you got your answer.

7 Get Real!

In America, lengths of fabric for making clothes are measured in yards and fractions of yards.
A tailor is making a suit for a customer.
The jacket needs $2\frac{1}{4}$ yards of fabric and the trousers need $1\frac{1}{3}$ yards.
The tailor has 4 yards of fabric.
How much will be left over when he has made the jacket and the trousers?

Explore

A man was riding a camel across a desert, when he came across three young men arguing. Their father had died, leaving seventeen camels as his sons' inheritance. The eldest son was to receive half of the camels; the second son, one-third of the camels and the youngest son, one-ninth of the camels. The sons asked him how they could divide seventeen camels in this way.

The man added his camel to the 17. Then, he gave $\frac{1}{2}$ of the camels to the eldest son, $\frac{1}{3}$ of the camels to the second son and $\frac{1}{9}$ of the camels to the youngest son. Having solved the problem, the stranger mounted his own camel and rode away.

How does this work?

(**Investigate further**)

Learn 6 Multiplying and dividing fractions

Examples:

a Calculate:

i $12 \times \frac{1}{3}$ **ii** $12 \times \frac{2}{3}$ **iii** $\frac{3}{4} \times \frac{2}{3}$

i $12 \times \frac{1}{3}$ ⟵——— Multiplying by $\frac{1}{3}$ is the same as dividing by 3

$= \frac{12}{3}$

$= 4$

ii $12 \times \frac{2}{3}$ ⟵——— Multiplying by $\frac{2}{3}$ is the same as dividing by 3 and multiplying by 2

$= \frac{24}{3}$

$= 8$

iii $\frac{3}{4} \times \frac{2}{3}$

$= \frac{3 \times 2}{4 \times 3}$

$= \frac{6}{12}$

$= \frac{1}{2}$

b Calculate:

i $8 \div \frac{1}{3}$ **ii** $8 \div \frac{2}{3}$ **iii** $\frac{3}{4} \div \frac{2}{3}$

i $8 \div \frac{1}{3}$ ⟵——— Dividing 8 by a third means finding how many thirds there are in 8
There are three thirds in each whole, so there are 8×3 thirds in 8

$= 8 \times \frac{3}{1}$

$= 24$

ii $8 \div \frac{2}{3}$ ⟵——— The number of two-thirds in 8 is half the number of thirds in 8
Dividing by $\frac{2}{3}$ is the same as multiplying by $\frac{3}{2}$

$= 8 \times \frac{3}{2}$

$= 12$

iii $\frac{3}{4} \div \frac{2}{3}$

$= \frac{3}{4} \times \frac{3}{2}$ ⟵——— Dividing by a fraction is the same as multiplying by the reciprocal (upside-down) fraction

$= \frac{3 \times 3}{4 \times 2}$

$= \frac{9}{8}$

$= 1\frac{1}{8}$

Apply 6

1 Work out:

a $18 \times \frac{1}{3}$ **c** $35 \times \frac{1}{7}$ **e** $40 \times \frac{2}{5}$ **g** $24 \times \frac{5}{8}$

b $28 \times \frac{1}{4}$ **d** $40 \times \frac{1}{5}$ **f** $12 \times \frac{3}{4}$ **h** $42 \times \frac{5}{6}$

2 Work out:

a $\frac{4}{5} \times \frac{1}{2}$ **c** $\frac{5}{6} \times \frac{1}{4}$ **e** $\frac{5}{8} \times \frac{3}{5}$ **g** $\frac{11}{12} \times \frac{4}{5}$

b $\frac{3}{8} \times \frac{1}{3}$ **d** $\frac{9}{10} \times \frac{1}{6}$ **f** $\frac{8}{9} \times \frac{3}{4}$ **h** $\frac{9}{10} \times \frac{2}{3}$

3 Work out:

a $18 \div \frac{1}{3}$ **c** $6 \div \frac{1}{5}$ **e** $18 \div \frac{2}{3}$ **g** $28 \div \frac{4}{5}$

b $5 \div \frac{1}{4}$ **d** $10 \div \frac{1}{8}$ **f** $12 \div \frac{3}{4}$ **h** $35 \div \frac{5}{6}$

4 Work out:

a $\frac{7}{8} \div \frac{1}{2}$ **c** $\frac{4}{9} \div \frac{2}{5}$ **e** $\frac{1}{3} \div \frac{1}{3}$ **g** $\frac{3}{5} \div \frac{7}{10}$

b $\frac{1}{6} \div \frac{2}{3}$ **d** $\frac{2}{7} \div \frac{2}{3}$ **f** $\frac{1}{3} \div \frac{1}{5}$ **h** $\frac{11}{12} \div \frac{3}{4}$

5 Paula is working out $\frac{4}{15} \div \frac{3}{8}$

She says, 'I can cancel the 4 into the 8 and the 3 into the 15.'

Then she writes down $\frac{1}{5} \div \frac{1}{2} = \frac{1}{5} \times \frac{2}{1} = \frac{2}{5}$

Is this correct? Explain your answer.

6 Ali says, 'This is how to divide fractions: $\frac{5}{6} \div \frac{2}{5} = \frac{6}{5} \times \frac{2}{5} = \frac{12}{25}$,'

Is Ali right? Explain your answer.

7 Without working out the answers, say which of these gives an answer greater than 1:

$\frac{9}{10} \times \frac{4}{5}$ or $\frac{9}{10} \div \frac{4}{5}$?

Give a reason for your answer.

8 Write down two fractions that:

a multiply to give 1   **b** divide to give 1.

9 Get Real!

Two thirds of the teachers in a school are women and three quarters of these are over 40. What fraction of the teachers in the school are women over 40?

10 Get Real!

Seven eighths of the members of the running club train on Wednesday evening and four fifths of them are male. What fraction of the members are males who train on Wednesday evenings?

Fractions

The following exercise tests your understanding of this chapter, with the questions appearing in order of increasing difficulty.

1 a Find the fraction that is shaded in each of the following diagrams. Write each fraction in its simplest form.

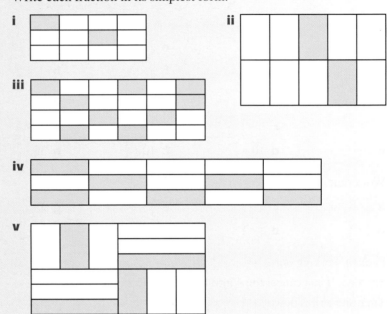

b Copy the following fractions and fill in the gaps.

i $\dfrac{3}{5} = \dfrac{\square}{10}$

iv $\dfrac{10}{12} = \dfrac{5}{\square}$

vii $\dfrac{9}{\square} = \dfrac{45}{100}$

ii $\dfrac{3}{4} = \dfrac{\square}{8}$

v $\dfrac{\square}{24} = \dfrac{5}{8}$

viii $\dfrac{8}{\square} = \dfrac{32}{60}$

iii $\dfrac{14}{24} = \dfrac{7}{\square}$

vi $\dfrac{\square}{12} = \dfrac{21}{36}$

2 a Change these fractions to sixteenths and then put them in order of size, starting with the smallest.

$\dfrac{3}{8}$ $\dfrac{1}{4}$ $\dfrac{7}{8}$ $\dfrac{9}{16}$

b Arrange these fractions in order of size, smallest first.

$\dfrac{2}{3}$ $\dfrac{5}{8}$ $\dfrac{7}{12}$ $\dfrac{11}{24}$

c Which is smaller, $\dfrac{2}{5}$ or $\dfrac{3}{8}$?

3 a Paula is running a 10 000 m race.
How far has she run when she has covered $\frac{5}{8}$ of it?

b Sunita spent $2\frac{1}{2}$ hours on her homework.
She spent $\frac{2}{5}$ of the time on her maths.
How long did she spend on maths?

c A bag contains 40 litres of plant food.
Dev uses $\frac{5}{8}$ of it in his patio tubs.
How many litres of plant food does he use?

d Mr Graham's school has 300 students.
$\frac{4}{15}$ of the students play musical instruments.
How many students play instruments?

e In a sale the price of a coat, originally £84, was reduced by $\frac{1}{3}$
What is the sale price of the coat?

4 a Mrs Snow travelled 480 miles to Scotland.
360 miles were on motorways.
What fraction of her journey was on motorways?

b A Virgin Voyager arrived at Euston station 30 minutes late.
What fraction is this of the timetabled journey time of $2\frac{1}{2}$ hours?

c What fraction of the word MISSISSIPPI is made up by I's?

d Old Macdonald had a farm.
He had 18 pigs, 5 goats, 128 sheep, 6 horses,
22 chickens, 20 cows and 1 bull.
What fraction of his livestock are sheep?

e In an examination sat by 240 students only 100 answered
a particular question correctly.
What fraction got the question wrong?

5 a Ann eats three quarters of a pound of fruit each day.
How much fruit does she eat in seven days?

b Tony eats $\frac{1}{4}$ kg of fruit each day.
He has 3 kg of fruit.
How many days will it take him to eat his fruit?

6 a Scrooge collects money.
$\frac{3}{10}$ of his fortune is in bronze coins, $\frac{8}{15}$ is in silver and the rest is in notes.
What fraction of Scrooge's fortune is in notes?

b In the 4×100 m relay, the first runner took $\frac{1}{5}$ of his team's total time.
The second runner took $\frac{7}{30}$ of their total time.
The third runner took $\frac{3}{10}$ of their total time.

  i What fraction of their time was taken by the fourth member of
the team?

  ii Which team member ran the fastest leg of the race?

  iii Which team member ran the slowest leg of the race?

c Delia is cooking.

She has a $1\frac{1}{2}$ kg bag of flour and needs $\frac{3}{8}$ of it in a recipe.

What fraction of a kilogram does she need and what is this in grams?

d S. Crumpy has an orchard.

The orchard contains $4\frac{1}{3}$ hectares of apple trees.

Today he needs to treat $\frac{4}{5}$ of the area for disease prevention.

What area does he need to treat?

e In a football match the goalkeeper kicked the ball from the goal line for $\frac{5}{8}$ of the length of the pitch and a player then kicked it a further $\frac{5}{24}$

The length of the pitch is 90 yards. How far is the ball from the opposing goal line?

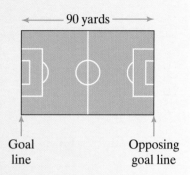

7 a Work out the following:

  i $6\frac{3}{7} + 3\frac{6}{7}$ **iii** $9\frac{4}{5} + 6\frac{3}{8}$ **v** $11\frac{2}{7} - 6\frac{4}{5}$ **vii** $10 \div \frac{2}{3}$

  ii $8\frac{1}{4} - 4\frac{5}{8}$ **iv** $7\frac{5}{8} - 3\frac{1}{4}$ **vi** $\frac{2}{3} \times \frac{5}{6}$ **viii** $3\frac{1}{5} \div 1\frac{3}{5}$

b i Titus Lines, the fisherman, catches one fish of mass $2\frac{1}{3}$ kg and another of mass $3\frac{1}{4}$ kg. What total mass of fish does he catch?

  ii What is the perimeter of a triangle of sides $2\frac{1}{4}$, $3\frac{1}{5}$ and $4\frac{3}{10}$ inches?

  iii A can holds $2\frac{8}{9}$ litres of oil. Hakim uses $1\frac{4}{15}$ litres. How much is left?

  iv Deirdre drops Ken off at work after driving from home for $4\frac{5}{12}$ miles. She drives $7\frac{1}{4}$ miles altogether to her own place of work. How far is Ken's workplace from Deirdre's workplace?

  v Milo is $1\frac{2}{7}$ metres tall. He is $\frac{3}{8}$ metre taller than Fizz. How tall is Fizz?

10 Representing data

OBJECTIVES

G

Examiners would normally expect students who get a G grade to be able to:

Construct and interpret a pictogram

Construct and interpret a bar chart

F

Examiners would normally expect students who get an F grade also to be able to:

Construct and interpret a dual bar chart

Interpret a pie chart

E

Examiners would normally expect students who get an E grade also to be able to:

Construct a pie chart

Interpret a stem-and-leaf diagram

D

Examiners would normally expect students who get a D grade also to be able to:

Construct a stem-and-leaf diagram (ordered)

Construct a frequency diagram

Interpret a time series graph

What you should already know ...

■ Measures of average including mean, median and mode

■ Accurate use of ruler and protractor

VOCABULARY

Bar chart – in a bar chart, the frequency is shown by the height (or length) of the bars. Bar charts can be vertical or horizontal

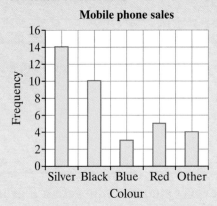

Mobile phone sales

Pictogram – in a pictogram, the frequency is shown by a number of identical pictures

Mobile phone sales

Colour	Frequency
Silver	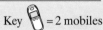
Black	
Blue	
Red	
Other	

Key = 2 mobiles

Pie chart – in a pie chart, frequency is shown by the angles (or areas) of the sectors of a circle

Mobile phone sales

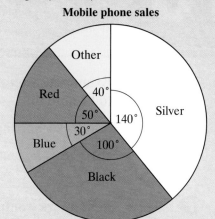

Frequency diagram – a frequency diagram is similar to a bar chart except that it is used for continuous data. In this case, there are usually no gaps between the bars

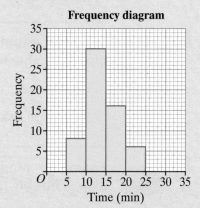

Frequency diagram

Stem-and-leaf diagram – a way of arranging data using a key to explain the 'stem' and 'leaf' so that $3\,|\,4$ represents 34

Number of minutes to complete a task

Stem (tens)	Leaf (units)
1	6 8 1 9 7
2	7 8 2 7 7 2 9
3	4 1 6

Key : $3\,|\,4$ represents 34 minutes

Ordered stem-and-leaf diagram – a stem-and-leaf diagram where the data is placed in order

Number of minutes to complete a task

Stem (tens)	Leaf (units)
1	1 6 7 8 9
2	2 2 7 7 7 8 9
3	1 4 6

Key : 3|4 represents 34 minutes

Back-to-back stem-and-leaf diagram – a stem-and-leaf diagram used to represent two sets of data

Number of minutes to complete a task

Leaf (units) Girls	Stem (tens)	Leaf (units) Boys
7 7 6 5 4 2 2	1	1 6 7 8 9
7 6 4 3 2 1	2	2 2 7 7 7 8 9
7 0	3	1 4 6

Key 3|2 represents 23 minutes

Key : 3|4 represents 34 minutes

Line graph – a line graph is a series of points joined with straight lines

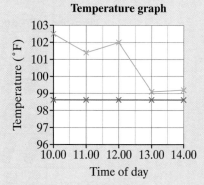

Temperature graph

Time series – a graph of data recorded at regular intervals

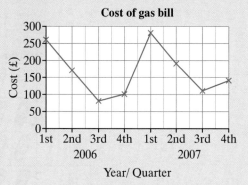

Cost of gas bill

Learn 1 Pictograms, bar charts and pie charts

Examples:

The following distribution shows the sales of mobile phones by colour.

Colour	Silver	Black	Blue	Red	Other
Frequency	14	10	3	5	4

Show this information as:

a a pictogram
b a bar chart
c a pie chart.

a In a pictogram, the frequency is shown by a number of identical pictures.

Mobile phone sales

Colour	Frequency
Silver	📱📱📱📱📱📱📱📱📱📱📱📱📱📱
Black	📱📱📱📱📱📱📱📱📱📱
Blue	📱📱📱
Red	📱📱📱📱📱
Other	📱📱📱📱

Key 📱 = 1 mobile

Mobile phone sales

Colour	Frequency
Silver	📱📱📱📱📱📱📱
Black	📱📱📱📱📱
Blue	📱📱
Red	📱📱📱
Other	📱📱

A key is used to say what each identical picture represents
Remember to provide a title to the work and include a key

Key 📱 = 2 mobiles

b Bar charts can be vertical or horizontal.

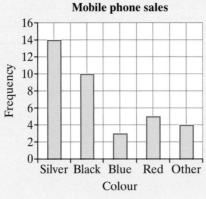

Mobile phone sales

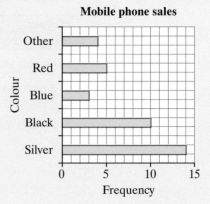

Mobile phone sales

Remember to label both axes and provide a title to the work
Use an appropriate scale to make the best use of the available space

c In a pie chart, the frequency is represented by the angles (or areas) of the sectors of a circle.

The pie chart needs to be drawn to represent 36 phones. There are 360° in a full circle, so each phone will be shown by 360° ÷ 36 = 10°

Colour	Frequency	
Silver	14	14 × 10° = 140°
Black	10	10 × 10° = 100°
Blue	3	3 × 10° = 30°
Red	5	5 × 10° = 50°
Other	4	4 × 10° = 40°
Total	36	360°

Draw your pie chart as large as possible and remember to label the sectors or provide a key.

Check that the sum of the angles does add up to 360°

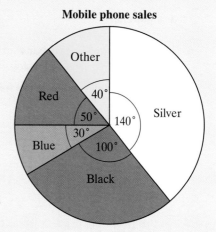

Mobile phone sales

Apply 1

1 The pictogram shows the number of ice creams sold over five days.

Day	Number of ice creams sold
Monday	🍦 🍦 🍦 🍦 🍦
Tuesday	🍦 🍦 🍦 🍦 🍦 🍦 🍦 🍦
Wednesday	
Thursday	🍦 🍦 🍦 🍦
Friday	🍦 🍦 🍦 🍦 🍦 🍦

Key: 🍦 = 4 ice creams

a How many ice creams were sold on Monday?

b How many ice creams were sold on Friday?

c On what day were most ice creams sold?

d What is the mean of the number of ice creams sold over the five days?

e How many ice creams were sold on Wednesday?

f Give a possible reason for your answer in part **e**.

2 The table shows the sales of fruit juices.

Fruit juice	Orange	Apple	Cranberry	Blackcurrant	Other
Frequency	18	8	3	6	1

Show this information as:

a a pictogram   **b** a bar chart   **c** a pie chart.

125

3 The table shows how 180 students travelled to college.

Travel	Car	Bus	Taxi	Walk	Cycle
Frequency	88	22	7	40	23

Show this information as:

a a bar chart **b** a pie chart.

4 The graph shows a dual bar chart for the number of e-mails and faxes received on five days of a week.

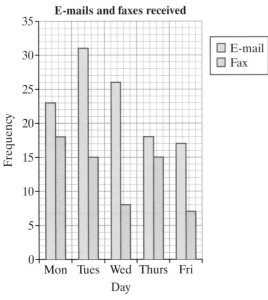

a Copy and complete the following table for the number of e-mails and faxes received.

Day	E-mails	Faxes
Monday		
Tuesday		
Wednesday		
Thursday		
Friday		

b What is the modal number of e-mails?

c What is the modal number of faxes?

d Calculate the range for the number of e-mails.

e Which day had the least number of faxes?

5 The table shows the gas and electricity bills (in pounds to the nearest pound) for the quarters in one year.

Quarter	Gas	Electricity
1st quarter	85	93
2nd quarter	52	68
3rd quarter	15	47
4th quarter	44	61

a Show this information in a dual bar chart.

b Use your graph to make five different comments about the gas and electricity bills.

6 The table shows the results of a survey to find students' favourite colours.
Draw a pie chart to show the information.

Colour	Tally
Red	|||| |||| ||||
Blue	|||| |||| |||| ||||	
Green	|||| ||||
Yellow	|||| |||
Black	||

7 Students at a college were asked to choose their favourite film.
Their choices are shown in the pie chart.

Favourite film

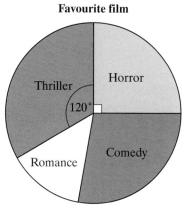

A total of 135 students chose horror films.

a How many students were included in the survey?

b How many students chose thrillers?

Twice as many students chose 'Comedy' as chose 'Romance'.

c How many students chose 'Comedy'?

8 Write down one advantage and one disadvantage for each of the
following representations of data.

a Pictogram

b Bar chart

c Pie chart

Learn 2 Stem-and-leaf diagrams

Examples:

a The number of minutes taken to complete an exercise was recorded for 15 boys in a class.

16, 27, 28, 22, 34, 18, 11, 19, 27, 31, 27, 36, 22, 17, 29

Show the information as a stem-and-leaf diagram.

Number of minutes to complete a task

Stem (tens)	Leaf (units)
1	6 8 1 9 7
2	7 8 2 7 7 2 9
3	4 1 6

In this case the number 6 stands for 16 (1 ten and 6 units)

In this case the number 6 stands for 36 (3 tens and 6 units)

Key: 3|4 represents 34 minutes

It is useful to provide an ordered stem-and-leaf diagram to find the median and range.

Number of minutes to complete a task

Stem (tens)	Leaf (units)
1	1 6 7 8 9
2	2 2 7 7 7 8 9
3	1 4 6

Here the leaves (units) are all arranged numerically

Key: 3|4 represents 34 minutes

b The number of minutes taken to complete an exercise was recorded for 15 boys and 15 girls in a class.

Boys: 16, 27, 28, 22, 34, 18, 11, 19, 27, 31, 27, 36, 22, 17, 29
Girls: 12, 23, 22, 17, 30, 16, 15, 14, 17, 37, 26, 24, 21, 12, 27

Show the information as a back-to-back stem-and-leaf diagram.

Here the leaves (units) are all arranged numerically from the right-hand side

Number of minutes to complete a task

Leaf (units) Girls	Stem (tens)	Leaf (units) Boys
7 7 6 5 4 2 2	1	1 6 7 8 9
7 6 4 3 2 1	2	2 2 7 7 7 8 9
7 0	3	1 4 6

Key: 3|2 represents 23 minutes

Key: 3|2 represents 32 minutes

The boys' data has already been recorded in an ordered stem-and-leaf diagram

This diagram is called a back-to-back (ordered) stem-and-leaf diagram.

Apply 2

1 The prices paid for some takeaway food are shown below.

£3.64 £4.15 £5.22 £5.88 £4.21 £4.55 £3.75 £4.78 £5.05 £4.52 £4.60

Copy and complete the following stem-and-leaf diagram to show this information.

Prices paid for takeaway food

Stem (£)	Leaf (pence)
3	
4	
5	

Key: 4|25 represents £4.25

2 The marks obtained in a test were recorded as follows.

8 20 9 21 18 22 19 13 22 24 14 9 25 10 19 20 17 14 12

a Show this information in an ordered stem-and-leaf diagram.

b What was the highest mark in the test?

c Write down the median of the marks in the test.

d Write down the range of the marks in the test.

3 The times taken to complete an exam paper were:

2 hr 12 min, 1 hr 53 min, 1 hr 26 min, 2 hr 26 min, 1 hr 50 min,
1 hr 46 min, 2 hr 05 min, 1 hr 43 min, 1 hr 49 min, 2 hr 10 min,
1 hr 49 min, 1 hr 55 min, 2 hr 06 min, 1 hr 57 min.

Show this information in an ordered stem-and-leaf diagram.

4 The heights of some students are shown in this stem-and-leaf diagram.

Heights of students

Stem (feet)	Leaf (inches)
4	10 09
5	08 07 03 11 08 01 10 00 08 07 06
6	02 01

Key: 5|06 represents 5 feet 6 inches

> **HINT** There are 12 inches in 1 foot.

a Rearrange the diagram to produce an ordered stem-and-leaf diagram.

b Use your diagram to answer these questions.

i How many students were included altogether?

ii Find the mode.

iii Find the median.

iv Calculate the range.

5 The number of minutes taken to complete an exercise was recorded for 15 boys and 15 girls in this back-to-back stem-and-leaf diagram.

Number of minutes to complete a task

Leaf (units) Girls	Stem (tens)	Leaf (units) Boys
7 7 6 5 4 2 2	1	1 6 7 8 9
7 6 4 3 2 1	2	2 2 7 7 7 8 9
7 0	3	1 4 6

Key: 3|2 represents 23 minutes   Key: 3|4 represents 34 minutes

a Calculate the median for the girls.

b Calculate the mode for the boys.

c Calculate the mean for the girls.

d Calculate the mean for the boys.

e Calculate the range for the girls.

f Calculate the range for the boys.

g What can you say about the length of time to complete the exercise by the girls and the boys?

6 Jenny records the reaction times of students in Year 7 and Year 11 at her school.

Year	Times (tenths of a second)
7	18 19 09 28 10 04 11 14 15 18 09 27 28 06 05
11	07 20 09 12 21 17 11 12 15 08 09 12 08 16 19

a Show this information in a back-to-back stem-and-leaf diagram.

b Jenny's hypothesis is 'The reaction times of Year 7 students are quicker than that of Year 11 students.'
Use your data to check Jenny's hypothesis.

Learn 3   Frequency diagrams and line graphs

Examples:

a 50 people were asked how long they had to wait for a train. The table below shows the results.

Time, *t* (minutes)	Frequency
$5 \leqslant t < 10$	16
$10 \leqslant t < 15$	22
$15 \leqslant t < 20$	11
$20 \leqslant t < 25$	1

Draw a frequency diagram to represent the data.

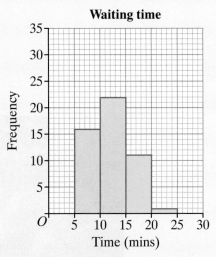

A frequency diagram is similar to a bar chart except that it is used for continuous data

b The table shows the temperature of a patient at different times during the day.

Time	10.00	11.00	12.00	13.00	14.00
Temperature (°F)	102.5	101.3	102	99.1	99.2

Draw a line graph to show this information.

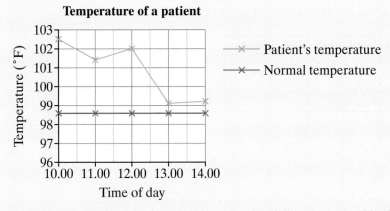

In this example, the temperature is expected to fall to the normal body temperature of 98.8°F. However, other line graphs can fluctuate over time.

Apply 3

1 The table shows the time spent in a local shop by 60 customers.

Time, t (minutes)	Frequency
$5 \leqslant t < 10$	8
$10 \leqslant t < 15$	30
$15 \leqslant t < 20$	16
$20 \leqslant t < 25$	6

Draw a frequency diagram to represent the data.

2 The frequency diagram shows the ages of 80 people in a factory.

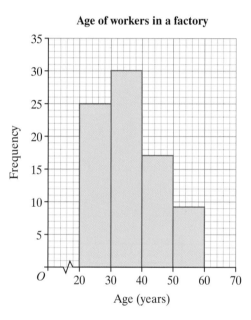

Age of workers in a factory

Copy and complete this table to show this information.

Age, y (years)	Frequency
$20 \leqslant y < 30$	
$30 \leqslant y < 40$	
$40 \leqslant y < 50$	
$50 \leqslant y < 60$	

3 The table shows the pressure in millibars (mb) over five days at a seaside resort.

Day	Pressure (mb)
Monday	1018
Tuesday	1022
Wednesday	1028
Thursday	1023
Friday	1019

Draw a line graph to show the pressure each day.

4 The table shows the minimum and maximum temperatures at a seaside resort.

Day	Minimum temperature (°C)	Maximum temperature (°C)
Monday	15	19
Tuesday	11	21
Wednesday	13	22
Thursday	14	23
Friday	17	23

Draw a line graph to show:

a the minimum temperatures.

b the maximum temperatures.

Use your graph to find:

c the day on which the lowest temperature was recorded

d the day on which the highest temperature was recorded

e the biggest difference between the daily minimum and maximum temperatures.

5 The table shows the cost of electricity bills at the end of every three months.

Year	2006	2006	2006	2006	2007	2007	2007	2007
Quarter	1st	2nd	3rd	4th	1st	2nd	3rd	4th
Cost	£230	£120	£50	£80	£215	£120	£25	£55

a Show this information on a graph.

b What can you say about the trend?

6 The table shows the number of students at college present during morning and afternoon registration.

Day	Mon	Mon	Tue	Tue	Wed	Wed	Thu	Thu	Fri	Fri
Session	am	pm	am	pm	am	pm	am	pm	am	pm
Number	220	210	243	215	254	218	251	201	185	152

a Show this information on a graph.

b What can you say about the trend?

Explore

Collect old bills together and see how they change over time

Suggestions include:

⊚ Gas bills

⊚ Electricity bills

⊚ Water bills

⊚ Telephone/mobile bills

⊚ Pay slips

Investigate further

Representing data

ASSESS

The following exercise tests your understanding of this chapter, with the questions appearing in order of increasing difficulty.

1 a The pictogram below shows the number of DVDs sold in a shop during one morning.

Time period	Number of DVDs sold
9.00–9.30	⊚ ⊚
9.30–10.00	(
10.00–10.30	⊚ ⊚ ⊚ ⊚ (
10.30–11.00	⊚ ⊚
11.00–11.30	⊚
11.30–12.00	⊚ ⊚ (

Key: ⊚ = 2 DVDs sold

i Write down the number of DVDs sold during each time period.

ii What is the difference in DVDs sold between the busiest and quietest time periods?

b The bar chart below shows the number of people in a library at noon on different days in a week.

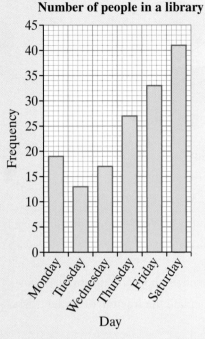

Number of people in a library

i Write down the number of people in the library at noon on each day.

ii What is the difference in the number of people in the library between the busiest and quietest days?

2 a The following information shows the number of flights from a regional airport on one day of the week made by different airlines.

Airline	Number of flights
Air OK	8
Flyme	9
Easyprop	4
Cyanair	3

Draw a pictogram to show this information.

b Five coins were tossed together 100 times and the number of tails showing uppermost was recorded as shown below.

Number of tails	Frequency
0	2
1	17
2	29
3	34
4	15
5	3

Draw a bar chart to show this information.

c The pie charts below show how 200 students travelled to college one day during winter and one day during summer.

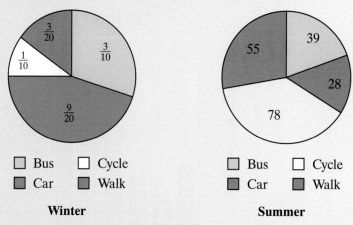

☐ Bus	☐ Cycle
■ Car	■ Walk

Winter

☐ Bus	☐ Cycle
■ Car	■ Walk

Summer

i Which method of transport was most popular on the winter day?

ii Calculate the number of students who travelled by bus on the winter day.

iii Which method of transport was most popular on the summer day?

iv Calculate the fraction of students who travelled by bus on the summer day.

3 a Becky keeps a note of all the first-class tickets she sells on the Eurostar service to Paris over a period of time.

Ticket	Frequency
Business	92
First Leisure	35
Child	17
Senior	24
Season ticket	12

Draw a pie chart to illustrate this information.

b There are 40 people in the choir at Beauvoice College. Their heights, to the nearest centimetres are shown in the stem-and-leaf diagram below.

Choir heights

Females		Males
9	14	
9 8 2	15	
9 9 8 7 6 6 6 4 4 3 1	16	2 4 7 9 9
8 7 5 3 3 2 2 1	17	2 2 4 5 5 8
2	18	3 3 6 9
	19	1

7|16 means 167 cm   18|2 means 182 cm

i How many females are in the choir?

ii What height is the tallest female?

iii What height is the shortest male?

iv How much taller is the tallest male than the shortest female?

c The number of students in a common room was counted at hourly intervals and the results recorded on the time series below.

Number of students in common room

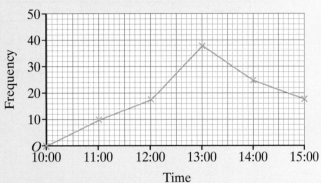

i How many students were in the common room at 11:00?

ii How many students were in the common room at 12:00?

iii How many students were in the common room at 14:30?

iv Why is your answer to part **iii** only an approximation?

v What is the maximum number of students recorded in the common room?

4 a The list shows the average gestation times, to the nearest day, of some animals. Show this data using a suitable stem-and-leaf diagram.

Animal	Gestation time (days)
Common opossum	13
Marine turtle	55
Grass lizard	42
Emperor penguin	63
House mouse	19
Royal albatross	79
Australian skink	30
Falcon	29
Hawk	44
Swan	30

Animal	Gestation time (days)
Python	61
Thrush	14
Wren	16
Spiny lizard	63
Alligator	61
Dog	63
Finch	12
Ostrich	42
Pheasant	22

b This data shows the track times in minutes and seconds on some of Luciano's CDs. Show the data using a suitable stem-and-leaf diagram.

3.21 2.29 2.25 2.49 2.57 3.30 3.19 2.25 3.34 2.45 2.44 3.34

3.19 3.30 2.10 3.00 2.44 2.25 2.54 3.43 2.22 2.54 2.55 2.24

3.35 2.10 3.55 3.07 2.54 2.08 3.22 3.33 2.43 3.50 2.22 2.57

Try a real past exam question to test your knowledge:

5 The table shows the number of people in the UK with a full time job.

Year	1980	1984	1988	1992	1996	2000
Number of people with a full time job (millions)	18.7	19.1	20.5	19.3	18.5	19.3

a Use the information to draw a time series graph.

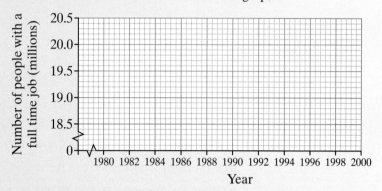

b Estimate the number of people in the UK with a full time job in 1997.

Spec B, Foundation Paper, Mar 02

11 Scatter graphs

OBJECTIVES

D **Examiners would normally expect students who get a D grade to be able to:**

Draw a scatter graph by plotting points on a graph

Interpret the scatter graph

Draw a line of best fit on the scatter graph

C **Examiners would normally expect students who get a C grade also to be able to:**

Interpret the line of best fit

Identify the type and strength of the correlation

What you should already know ...

■ Use coordinates to plot points on a graph

■ Draw graphs including labelling axes and giving a title

VOCABULARY

Coordinates – a system used to identify a point; an x-coordinate and a y-coordinate give the horizontal and vertical positions

Correlation – a measure of the relationship between two sets of data; correlation is measured in terms of type and strength

Strength of correlation

The strength of correlation is an indication of how close the points lie to a straight line (perfect correlation)

Strong correlation   **Weak correlation**

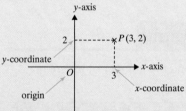

Correlation is usually described in terms of strong correlation, weak correlation or no correlation

Type of correlation

Positive correlation   **Negative correlation**

In positive correlation an increase in one set of variables results in an increase in the other set of variables

In negative correlation an increase in one set of variables results in a decrease in the other set of variables

Zero or no correlation

Zero or no correlation is where there is no obvious relationship between the two sets of data

Line of best fit – a line drawn to represent the relationship between two sets of data. Ideally it should only be drawn where the correlation is strong, for example,

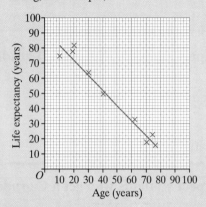

Scatter graph – a graph used to show the relationship between two sets of variables, for example, temperature and ice cream sales

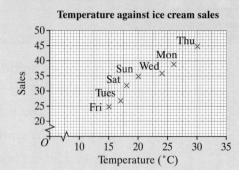

Outlier – a value that does not fit the general trend, for example,

Learn 1   Plotting points on a scatter graph

Example:

A shopkeeper notes the temperature and the number of ice creams sold each day.

	Sun	Mon	Tue	Wed	Thu	Fri	Sat
Temperature (°C)	20	26	17	24	30	15	18
Ice cream sales	35	39	27	36	45	25	32

Plot this information on a scatter graph.

The information can be plotted as a series of coordinate pairs.

	Sun	Mon	Tue	Wed	Thu	Fri	Sat
Temperature (°C)	20	26	17	24	30	15	18
Ice cream sales	35	39	27	36	45	25	32
	(20, 35)	(26, 39)	(17, 27)	(24, 36)	(30, 45)	(15, 25)	(18, 32)

Before drawing a scatter graph, you need to choose carefully the scales on the axes.

On the *y*-axis, each small square represents 0.1 units

On the *x*-axis, each small square represents 10 units

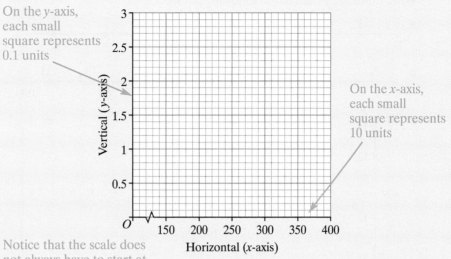

Notice that the scale does not always have to start at zero; a jagged line is often used to show the scale does not start at *O*

Temperature against ice cream sales

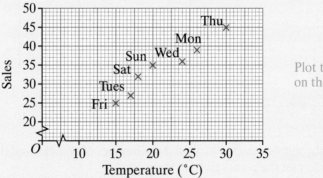

Plot the first variable on the horizontal axis

You can see from the graph that as the temperature rises, the sales of ice cream increase.

Apply 1

1 Write down the values of the following points.

a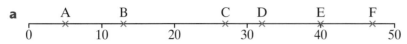

b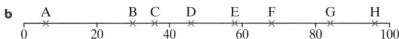

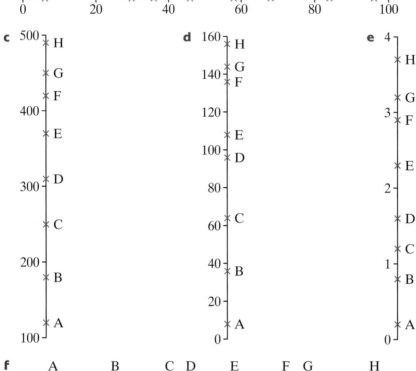

f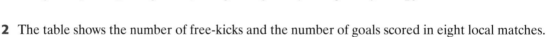

2 The table shows the number of free-kicks and the number of goals scored in eight local matches.
Copy the graph and plot the points given in the table.

Number of free-kicks	2	3	5	9	8	1	3	7
Number of goals	5	7	4	1	3	7	6	5

3 The table shows the ages of eight friends and their shoe sizes.
Copy the graph and plot the points given in the table.

Age (years)	4	9	7	13	6	8	11	12
Shoe size	2	5	6	9	3	5	8	8

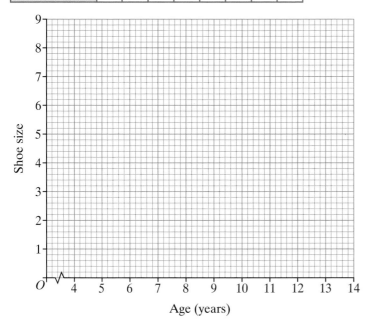

4 The table shows the number of spectators and the number of goals scored in eight school matches.

Copy the graph and plot the points given in the table.

Number of spectators	15	40	52	8	27	5	23	24
Number of goals	2	4	1	5	3	1	6	1

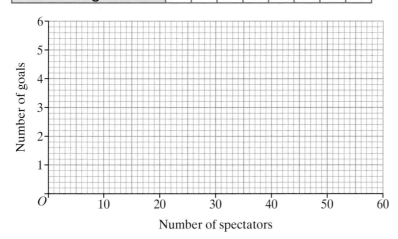

5 The table shows the weights of bikes (in kg) and their cost (in $).
Copy the graph and plot the points given in the table.

Weight of bike (kg)	6.7	11.4	8.2	7.1	9.0	10.2	8.1	9.6	11.2
Cost in dollars ($)	69	94	88	73	80	88	79	86	84

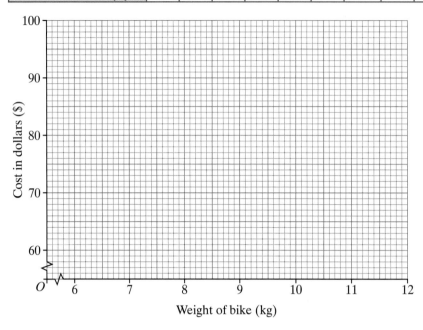

Weight of bike (kg)

6 The table shows the weights of runners (in kg) and their times to complete 100 m (in seconds).

Copy the graph and plot the points given in the table.

Weight of runner (kg)	64	110	85	72	94	104	77	97
Time (sec)	12.7	10.4	11.5	11.8	11.3	10.6	12.0	10.9

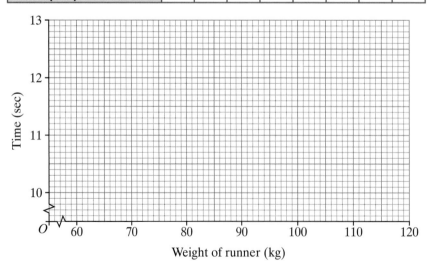

Weight of runner (kg)

7 Copy and complete the table below, using the graph to find your values.

Point	A	B	C	D	E	F	G	H	I	J
Days absent										
Number of people										

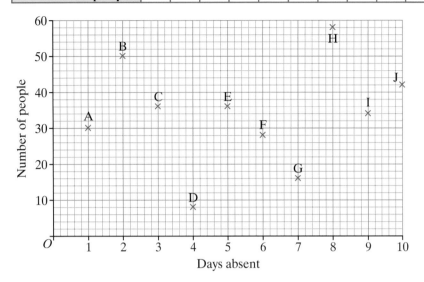

8 Copy and complete the table below, using the graph to find your values.

Point	A	B	C	D	E	F	G	H	I	J
Age (years)										
Amount earned (£ thousand)										

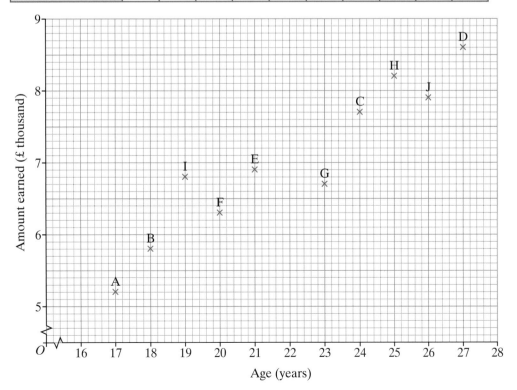

Learn 2   Interpreting scatter graphs

Example: Use the graph to describe the correlation between temperature and ice cream sales.

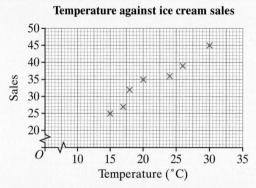

Temperature against ice cream sales

Correlation measures the relationship between two sets of data
It is measured in terms of the **strength** and **type** of correlation

You can see from the graph that as the temperature rises, the sales of ice cream increase. There is a link between the temperature and the sales of ice cream.

Therefore, there is **strong positive** correlation between the ice cream sales and the temperature.

Apply 2

1 For each of these scatter graphs:

  i   describe the type and strength of correlation

  ii   write a sentence explaining the relationship between the two sets of data (for example, the higher the rainfall, the heavier the weight of apples).

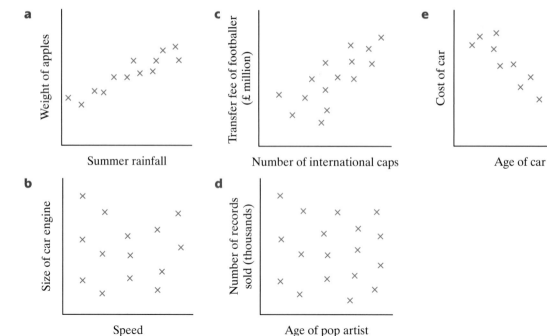

  a   Weight of apples / Summer rainfall

  b   Size of car engine / Speed

  c   Transfer fee of footballer (£ million) / Number of international caps

  d   Number of records sold (thousands) / Age of pop artist

  e   Cost of car / Age of car

2 For each of the following:

   i describe the type and strength of correlation

   ii write a sentence explaining the relationship between the two sets of data.

   a The hours of sunshine and the income from hiring deckchairs.

   b The number of cars on the road and the average speed of cars

   c The distance travelled and the money spent on petrol.

   d The height of a person and the number of children in his or her family.

3 The table shows the ages and arm spans of seven students in a school.

Age (years)	16	13	13	10	18	10	15
Arm span (inches)	62	57	59	57	64	55	61

   a Draw a scatter graph of the results.

   b Describe the type and strength of correlation.

   c Write a sentence explaining the relationship between the two sets of data.

4 The table shows the ages and second-hand values of seven cars.

Age of car (years)	2	1	4	7	10	9	8
Value of car (£)	4200	4700	2800	1900	400	1100	2100

   a Draw a scatter graph of the results.

   b Describe the type and strength of correlation.

   c Write a sentence explaining the relationship between the two sets of data.

5 The table shows the daily rainfall and the number of sunbeds sold at a resort on the south coast.

Amount of rainfall (mm)	0	1	2	5	6	9	11
Number of sunbeds sold	380	320	340	210	220	110	60

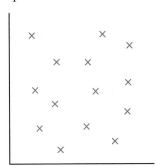

   a Draw a scatter graph of the results.

   b Describe the type and strength of correlation.

   c Write a sentence explaining the relationship between the two sets of data.

6 For each graph, write down two variables that might fit the relationship.

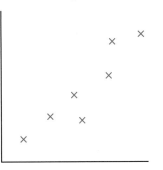

7 The scatter graph shows the ages and shoe sizes of a group of people.

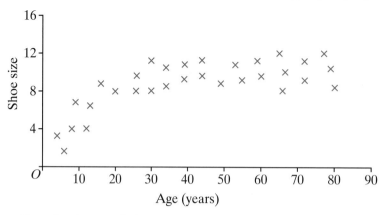

a Describe the correlation.

b Give a reason for your answer.

8 Get Real!

Steve is investigating the fat content and the calorie values of food at his local fast-food restaurant.
He collects the following information.

	Fat (g)	Calories
Hamburger	9	260
Cheeseburger	12	310
Chicken nuggets	24	420
Fish sandwich	18	400
Medium fries	16	350
Medium cola	0	210
Milkshake	26	1100
Breakfast	46	730

a Draw a scatter graph of the results.

b Describe the type and strength of correlation.

c Does the relationship hold for all the different food?
Give a reason for your answer.

Explore

◎ Undertake some research of your own into the fat content and calorie values for food in a local restaurant

◎ You can find this information from the restaurant itself or on the internet

Investigate further

Learn 3 Lines of best fit

Example: Use the graph to draw a line of best fit and estimate the likely sales of ice cream for a temperature of 28°C.

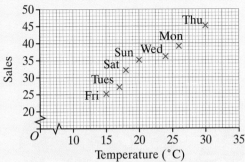

Temperature against ice cream sales

Drawing the line of best fit, you can use the graph to estimate the likely sales of ice creams for a temperature of 28°C.

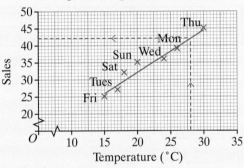

Temperature against ice cream sales

For some graphs a straight line is not possible and a curve of best fit may be appropriate

From the graph, the number of ice cream sales is 42.

However, care should be taken when making use of such methods. For example, you will notice that the sales of ice creams were slightly higher on Saturday and Sunday than the line would predict.

You might also use your line of best fit to estimate the temperature given the number of ice cream sales – but it is not likely that you would need to do this.

Apply 3

1 The table shows the ages and arm spans of seven students in a school.

Age (years)	16	13	13	10	18	10	15
Arm span (inches)	62	57	59	57	64	55	61

a Draw a line of best fit on the scatter graph you drew in Apply **2**, question **3**.

b Use your line of best fit to estimate:

i the arm span of an 11-year-old student

ii the age of a student with an arm span of 61 inches.

2 The table shows the ages and second-hand values of seven cars.

Age of car (years)	2	1	4	7	10	9	8
Value of car (£)	4200	4700	2800	1900	400	1100	2100

a Draw a line of best fit on the scatter graph you drew in Apply **2**, question **4**.

b Use your line of best fit to estimate:

i the value of a car if it is 7.5 years old

ii the age of a car if its value is £3700.

3 The table shows the daily rainfall and the number of sunbeds sold at a resort on the south coast.

Amount of rainfall (mm)	0	1	2	5	6	9	11
Number of sunbeds sold	380	320	340	210	220	110	60

a Draw a line of best fit on the scatter graph you drew in Apply **2**, question **5**.

b Use your line of best fit to estimate:

i the number of sunbeds sold for 4 mm of rainfall

ii the amount of rainfall if 100 sunbeds were sold.

4 Sally collects information on the temperature and the number of visitors to a museum.

Temperature (°C)	15	25	16	18	19	22	24	23	17	20	26	20
Number of visitors	720	180	160	620	510	400	310	670	720	530	180	420

a Draw a scatter graph and a line of best fit.

b Use your line of best fit to estimate:

i the number of visitors if the temperature is 24°C

ii the temperature if 350 people visit the museum.

c Sally is sure that two of the data pairs are incorrect. Identify these two pairs on your graph.

5 The graph shows the line of best fit for the relationship between house prices in 2000 and house prices in 2006.

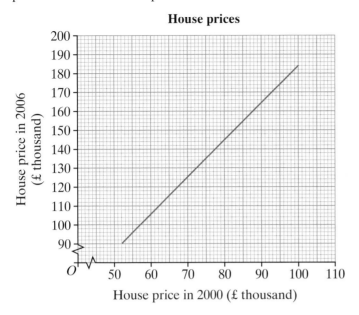

House prices

a Copy and complete the table, giving estimates for the missing values.

House price in 2000 (£ thousand)	70	75	90	100		
House price in 2006 (£ thousand)	125				105	155

b Andrea says her house price was £80 000 in 2000 and £170 000 in 2006. Is she correct? Give a reason for your answer.

c Find an estimate for the 2006 price of a house priced £55 000 in 2000.

d Find an estimate for the 2000 price of a house priced £180 000 in 2006.

6 Readings of two variables, *A* and *B*, are shown in the table.

A	1	2	3	4	5	6	7	0.8	2.1	3.2	3.9	5.1	6.2	7.1
B	1.8	8.8	20	33	48	73	95	2	9	18	31	49	72	98

a Draw a scatter graph.

b What can you say about the correlation between the two sets of data?

c Draw a curve of best fit and use this to estimate:

   i the value of *B* if $A = 2.5$

   ii the value of *A* if $B = 64$.

Explore

- ⊚ Investigate the cost and age of cars – you may wish to use the internet
- ⊚ What relationship do you notice?
  Are there any exceptions?

Investigate further

Explore

- ⊚ Investigate house prices and the number of bedrooms – you may wish to use a local newspaper
- ⊚ What relationship do you notice?
  Are there any exceptions?

Investigate further

Explore

- ⊚ Investigate body lengths – for example, your arm span and your height are closely related
- ⊚ Collect data from your friends and family to see what relationships you can find

Investigate further

Scatter graphs

The following exercise tests your understanding of this chapter, with the questions appearing in order of increasing difficulty.

1 The table shows the marks of eight students in English and mathematics.

Student	1	2	3	4	5	6	7	8
English	25	35	28	30	36	44	15	21
Mathematics	27	40	29	32	41	48	17	20

Draw a scatter graph and comment on the relationship between the marks in the two subjects.

2 Mr Metcalf, the maths teacher, told his class they had a test in a week's time. He also asked them to record how many hours of TV they watched during the week before the test. When he had marked the test he showed the class a scatter graph of the data in the table.

Student	1	2	3	4	5	6	7	8	9	10
TV watched (hours)	4	7	9	10	13	14	15	20	21	25
Test mark	92	90	74	30	74	66	95	38	35	30

a Draw a scatter graph and comment on the relationship between the marks in the test and the amount of TV watched.

b Two students do not seem to 'fit the trend'.
Identify the students and give a possible reason for their results.

3 The table shows the relationship between the area (in thousands of km^2) of some countries and their populations (in millions), all to 2 s.f.

	Area	Population
Monaco	0.0020	0.030
Malta	0.32	0.40
Jersey	0.12	0.090
Netherlands	42	16
UK	250	60
Germany	360	83
Italy	300	58
Switzerland	41	7.3
Andorra	0.47	0.068
Denmark	43	5.4
France	550	60
Austria	84	8.2
Turkey	780	67
Greece	130	11
Spain	500	40
Ireland	70	3.9
Latvia	65	2.4
Sweden	450	8.9
Norway	320	4.5
Iceland	100	0.28

Draw a scatter graph of this data and comment on the graph.

4 a Draw a suitable scatter graph to illustrate this data, which shows the relationship between the distances jumped in long jump trials and the leg lengths of the jumpers.

Leg length (cm)	71	73	74	75	76	79	82
Distance jumped (m)	3.2	3.1	3.3	4.1	3.9	4	4.8

b Draw a line of best fit on the graph.

c Use your line of best fit to estimate:

   i the leg length of an athlete who jumped a distance of 3.5 m

   ii the distance jumped by an athlete with a leg length of 85 cm.

d Explain why one of these estimates is more reliable than the other.

Try a real past exam question to test your knowledge:

5 The scatter graph shows the height and trunk diameter of each of eight trees.

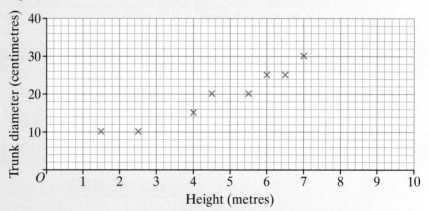

a What is the height of the tallest tree?

b Draw a line of best fit through the points on the scatter graph.

c Describe the relationship shown in the scatter graph.

d i Estimate the height of a tree with trunk diameter 35 centimetres.

   ii Comment on the reliability of your estimate.

Spec B, Int Paper 1, Mar 02

12 Properties of polygons

OBJECTIVES

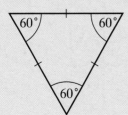

G **Examiners would normally expect students who get a G grade to be able to:**

Recognise and name shapes such as isosceles triangle, parallelogram, rhombus, trapezium and hexagon

E **Examiners would normally expect students who get an E grade also to be able to:**

Calculate interior and exterior angles of a quadrilateral

Investigate tessellations

C **Examiners would normally expect students who get a C grade also to be able to:**

Classify a quadrilateral by geometric properties

Calculate exterior and interior angles of a regular polygon

What you should already know ...

- Recall and use properties of angles at a point, angles on a straight line, perpendicular lines, and opposite angles at a vertex

- Distinguish between acute, obtuse, reflex and right angles

- Use parallel lines, alternate angles and corresponding angles

- Prove that the angle sum of a triangle is 180°

- Prove that the exterior angle of a triangle is equal to the sum of the interior opposite angles

- Use angle properties of equilateral, isosceles and right-angled triangles

- Understand simple congruence

VOCABULARY

Equilateral triangle – a triangle with 3 equal sides and 3 equal angles – each angle is 60°

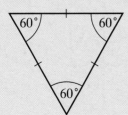

Isosceles triangle – a triangle with 2 equal sides and 2 equal angles; the equal angles are called **base angles**

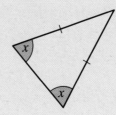

the x angles are base angles

Square – a quadrilateral with four equal sides and four right angles

Rectangle – a quadrilateral with four right angles, and opposite sides equal in length

Kite – a quadrilateral with two pairs of equal adjacent sides

Trapezium – a quadrilateral with one pair of parallel sides

Isosceles trapezium – a quadrilateral with one pair of parallel sides. Non-parallel sides are equal

Parallelogram – a quadrilateral with opposite sides equal and parallel

Rhombus – a quadrilateral with four equal sides and opposite sides parallel

Polygon – a closed two-dimensional shape made from straight lines

Pentagon – a polygon with five sides

Hexagon – a polygon with six sides

Heptagon – a polygon with seven sides

Octagon – a polygon with eight sides

Nonagon – a polygon with nine sides

Decagon – a polygon with ten sides

Regular polygon – a polygon with all sides and all angles equal

Irregular polygon – a polygon whose sides and angles are not all equal (they do not all have to be different)

Interior angle – an angle inside a polygon

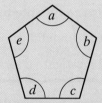 *a, b, c, d* and *e* are interior angles

Exterior angle – if you make a side of a triangle longer, the angle between this and the next side is an exterior angle

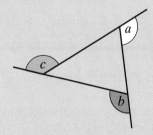 *a, b* and *c* are exterior angles

Convex polygon – a polygon with no interior reflex angles

Concave polygon – a polygon with at least one interior reflex angle

Tessellation – a pattern where one or more shapes are fitted together repeatedly leaving no gaps

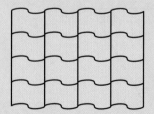

Learn 1   Angle properties of quadrilaterals

Example: Calculate the angles marked with letters in the shape below.

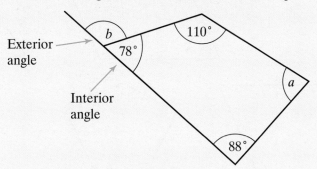

The angles in the quadrilateral add up to 360°, so

$a = 360° - (78° + 88° + 110°)$
$= 360° - 276°$
$= 84°$

The exterior and interior angles add up to 180°, so

$b = 180° - 78°$
$= 102°$

Apply 1

1 Name these quadrilaterals.

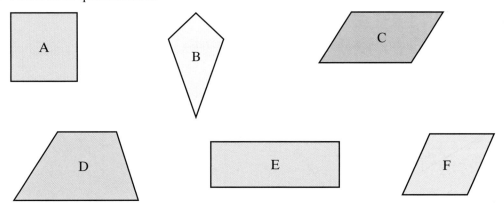

2 Calculate the size of the angles marked with letters.

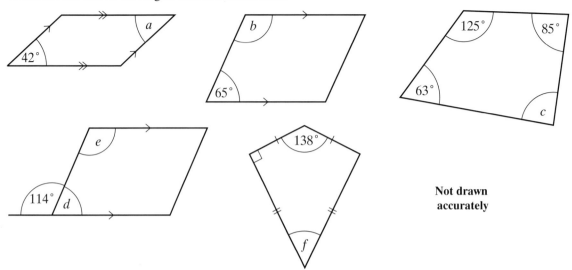

Not drawn accurately

3 a Three angles of a quadrilateral are 60°, 65° and 113°.
Find the size of the fourth angle.

   b Two angles of a quadrilateral are 74° and 116° and the other two angles
are equal. What size are the other two angles?

   c If all four angles of a quadrilateral are equal, what size are they?
What sort of quadrilateral could it be?

4 A quadrilateral has three angles of 84°.
What is the size of the fourth angle?

5 An isosceles trapezium has an angle of 83°.
What are the sizes of the other three angles?

6 Calculate the sizes of the marked angles in these quadrilaterals.

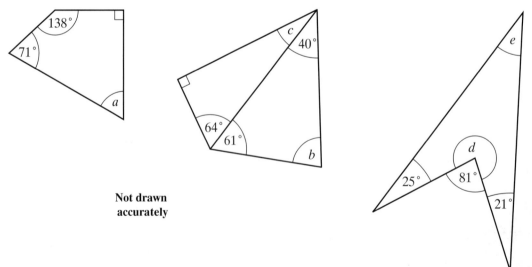

**Not drawn
accurately**

7 Barry measures the angles of a quadrilateral. He says that three of the
angles are 82° and the other one is 124°. Could he be right?
Explain your answer.

8 Harry measures the angles of a quadrilateral. He says that the angles are 72°, 66°, 114° and 108°. He says the shape is a trapezium. Could he be right? Explain your answer.

9 Larry measures the angles of a quadrilateral. He says that two of them are 78°, and two of them are 102°. What quadrilateral might he have been measuring? (There are three possible answers; can you find them all?)

10 Get Real!

Khadija has two pieces of card in the shape of equilateral triangles. Each side is 4 cm long.

She makes a puzzle for her brother like this.

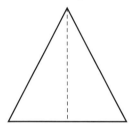 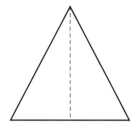 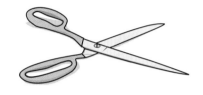

She cuts each piece of card in half along the dotted lines as shown. She now has four congruent right-angled triangles.

a She asks her brother to rearrange the four triangles to make:

  i a rectangle

  ii a trapezium

  iii a parallelogram

  iv a rhombus.

Draw a picture to show her brother's answers.

b Work out the size of the angles in one of the right-angled triangles.

11 Sally's teacher wants the class to draw a trapezium with two sides of 5 cm, one side of 7 cm and one side of 10 cm. Sally says there is more than one answer. Is she right? Give a reason for your answer.

Explore

◎ A kite always has an obtuse angle
   True or false?

◎ Can a kite have two obtuse angles?

Investigate further

Learn 2 Diagonal properties of quadrilaterals

Example: The diagonals of a quadrilateral are different lengths, but one is the perpendicular bisector of the other.
What is the mathematical name of the quadrilateral?

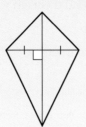

The quadrilateral is a kite.

Apply 2

1 a Copy these shapes and draw in their diagonals.

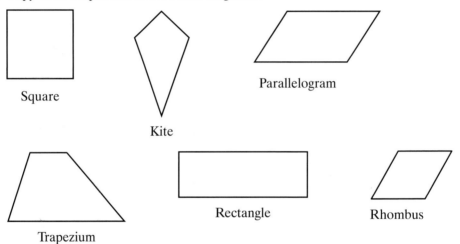

Square

Kite

Parallelogram

Trapezium

Rectangle

Rhombus

b Use your diagrams to complete a copy of this table.

Shape	Are the diagonals equal? (Yes/No)	Do the diagonals bisect each other? (Both/One only/No)	Do the diagonals cross at right angles? (Yes/No)	Do the diagonals bisect the angles of the quadrilateral? (Yes/Two only/No)
Square				
Kite				
Parallelogram				
Trapezium				
Rectangle				
Rhombus				

2 Rajesh says that he has drawn a quadrilateral. Its diagonals are equal.
What quadrilaterals might he have drawn?
(Use the table from question **1** to help you.)

3 Michelle says that the diagonals of a rectangle bisect the
corner angles.
So angles *a* and *c* are both 45°, and angle *b* must be 90°.
Is she right? Explain your answer.

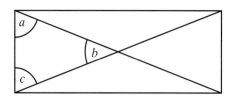

4 The diagram below shows a rhombus ABCD. AC and BD are the diagonals.
Angle ADB = 32°.

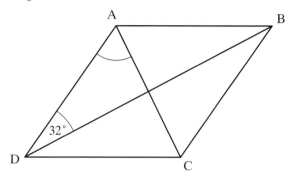

**Not drawn
accurately**

Calculate angle DAC.

5 Calculate the angles marked with letters in the diagrams below.
You will need to use the properties of diagonals.
Give a reason for each of the angles.

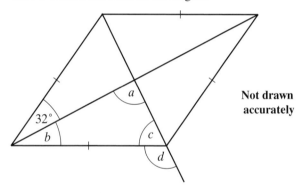

**Not drawn
accurately**

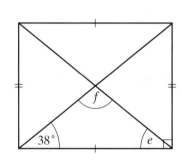

6 Get Real!
A builder has two types of wall tile. One is an isosceles trapezium,
the other is a rhombus. He creates this tessellation.

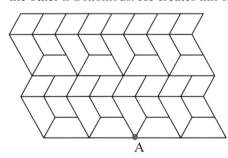

a By looking at the angles around point A, calculate the angles
of the trapezium.

b Now calculate the angles of the rhombus.

7 Copy and complete this table. The top line has been done for you.

Shape	Number of different length sides (at most)	Number of right angles (at least)	Pairs of opposite sides parallel	Diagonals must be equal	Diagonals bisect each other	Diagonals cross at right angles
Square	1	4	Both	Yes	Yes	Yes
Rectangle						
Trapezium						
Rhombus						
Parallelogram						
Kite						
Isosceles trapezium						

Explore

◎ Roger draws a quadrilateral
He uses it to make this tessellation pattern

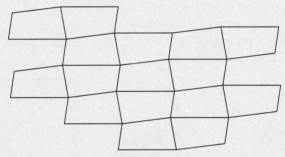

◎ He says you can make a tessellation with any quadrilateral

Investigate further

Learn 3 Angle properties of polygons

Examples:

a Calculate the sum of the interior angles of a:

i pentagon   **ii** hexagon.

i The pentagon can be divided into three triangles by drawing diagonals from a start point.

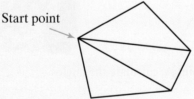

Start point

The sum of the angles is
$(5 - 2) \times 180° = 540°$

ii The hexagon can be divided into four triangles in the same way.

Start point

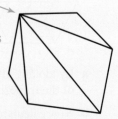

The interior angles of a polygon add up to (number of sides – 2) × 180°

The sum of the angles is
$(6 - 2) \times 180° = 720°$

b Calculate the interior angle of a regular octagon.

To calculate the interior angle of a regular octagon you can use one of the following methods.

Either:
An octagon has eight sides.
The sum of the angles can be found by dividing the octagon into six triangles as shown.
So the sum of the angles is $(8 - 2) \times 180° = 1080°$.
A regular octagon has all angles equal, so each angle is $1080° \div 8 = 135°$.

Start point

Or:
A regular octagon has eight equal exterior angles.
So each exterior angle is $360° \div 8 = 45°$.
So each interior angle is $180° - 45° = 135°$.

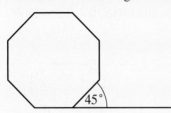

c A regular polygon has interior angles of 144°.
How many sides does it have?

The exterior angles of a convex polygon add up to 360°

A regular polygon has all sides equal and all angles equal
Always draw a diagram to help answer the questions. You can label the diagram to keep track of what you know

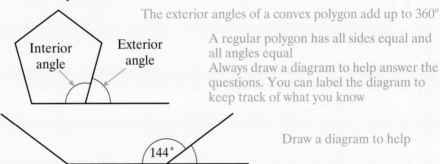

Draw a diagram to help

Each exterior angle must be $180° - 144° = 36°$.
Exterior angles add up to 360°, so there must be $360 \div 36 = 10$ exterior angles, and 10 sides.

Apply 3

1 Calculate the angles marked with letters in the diagram.
Explain how you worked them out.

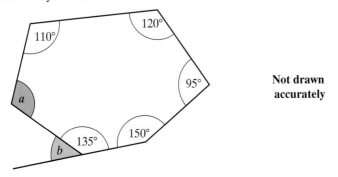

Not drawn accurately

2 Four of the angles of a pentagon are 110°, 130°, 102° and 97°.
Calculate the fifth angle.

3 a A regular polygon has an exterior angle of 40°. How many sides has it?

   b Will this polygon tessellate?

4 Some of these quadrilaterals are possible, and others are not.
If possible draw:

   a a kite with a right angle   **f** a triangle with two right angles

   b a kite with two right angles   **g** a pentagon with one right angle

   c a trapezium with two right angles   **h** a pentagon with two right angles

   d a trapezium with only one right angle   **i** a pentagon with three right angles

   e a triangle with a right angle   **j** a pentagon with four right angles.

5 Elena says that the interior angles of a decagon add up to $10 \times 180° = 1800°$.
Is she right? Give a reason for your answer.

6 Get Real!

A company makes containers as shown.
The top is in the shape of a
regular octagon.

   a What is the size of each interior angle?

   b When the company packs them into a box,
will they tessellate? If not, what shape will be
left between them?

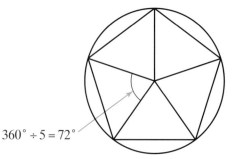

7 The diagrams show how you can draw an equilateral triangle and a regular
hexagon inside a circle by dividing the angle at the centre into equal parts.

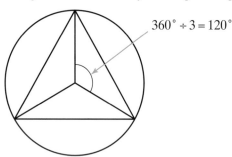

$360° \div 3 = 120°$

$360° \div 5 = 72°$

Use the same method to draw a regular hexagon and a regular nonagon
inside a circle.

8 The diagram shows a regular pentagon ABCDE and a regular hexagon DEFGHI.

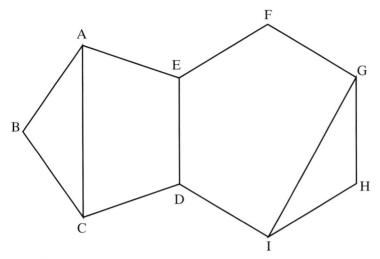

Calculate:

a angle EDC

b angle EDI

c obtuse angle CDI

d angle BAC

e angle CAE

f angle HIG

g angle DIG

Properties of polygons

The following exercise tests your understanding of this chapter, with the questions appearing in order of increasing difficulty.

1 a Name these shapes.

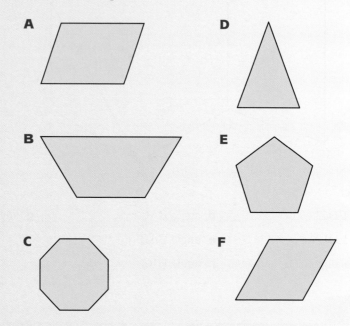

b Write down the letters of the diagrams that have:

i some sides equal

ii all sides equal

iii any acute angles

iv some equal angles

v any adjacent sides equal

vi all diagonals equal

vii diagonals perpendicular to each other

viii any adjacent angles equal.

2 a Will **any** parallelogram tessellate?
Explain your answer.

b Will **any** rhombus tessellate?
Explain your answer.

3 a Draw a quadrilateral that can be cut into two pieces by drawing a straight line across it.

b Draw a quadrilateral that can be cut into three pieces by drawing a straight line across it.

4 Find the values of the marked angles in these diagrams.

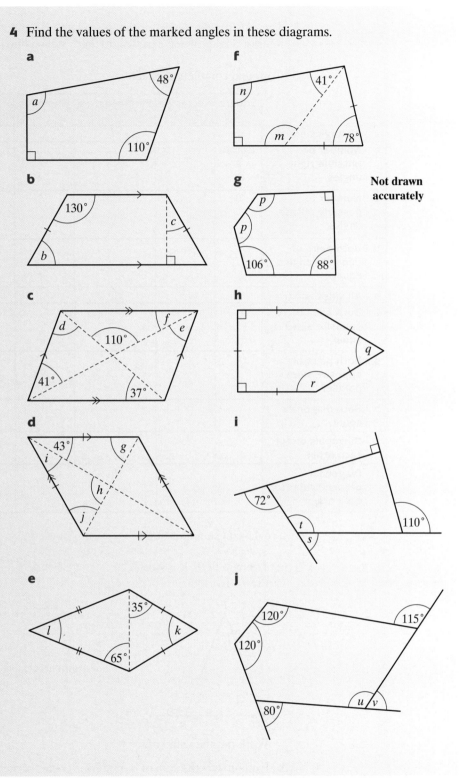

Not drawn accurately

5 The only regular polygons that tessellate on their own are those whose interior angles divide exactly into 360°. Which ones are they?

6 Copy and complete the table.

		Square	Rectangle	Parallelogram	Rhombus	Trapezium	Isosceles trapezium	Kite
a	All angles equal	Yes						
b	Number of possible right angles	Exactly 4				0 or 2		1 or 2
c	Number of possible obtuse angles	0		Exactly 2				
d	Both pairs of opposite angles equal		Yes					
e	All sides equal			No				
f	Both pairs of opposite sides equal					No		
g	Both pairs of opposite sides parallel						No	
h	Both diagonals equal							No
i	Diagonals bisect each other		Yes					
j	Diagonals perpendicular to each other			No				

Try a real past exam question to test your knowledge:

7 a Triangle PQR is isosceles.
PQ = PR.

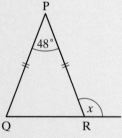

Not drawn accurately

Work out the value of x.

b Explain why the sum of the interior angles of any quadrilateral is 360°.

Spec A, Int Paper 2, Nov 04

13 Indices

OBJECTIVES

F ▶ **Examiners would normally expect students who get an F grade to be able to:**

Calculate squares and square roots (with and without the use of a calculator)

E ▶ **Examiners would normally expect students who get an E grade also to be able to:**

Calculate cubes and cube roots (with and without the use of a calculator)

Use function keys on a calculator for powers and roots

D ▶ **Examiners would normally expect students who get a D grade also to be able to:**

Use the terms square, positive square root, negative square root, cube and cube root

Recall integer squares from 2×2 to 15×15 and the corresponding square roots

Recall the cubes of 2, 3, 4, 5 and 10

C ▶ **Examiners would normally expect students who get a C grade also to be able to:**

Use index notation and index laws for positive and negative powers such as $w^3 \times w^5$ and $\dfrac{w^3}{w^7}$

What you should already know ...

- Understand the idea of square numbers
- Understand the idea of reciprocals
- Basic use of algebra

Square number – a square number is the outcome when a number is multiplied by itself

Cube number – a cube number is the outcome when a number is multiplied by itself then multiplied by itself again

Square root – a square root of a number such as 16 is a number whose outcome is 16 when multiplied by itself

Cube root – the cube root of a number such as 125 is a number whose outcome is 125 when multiplied by itself then multiplied by itself again

Index or **power** or **exponent** – the index tells you how many times the base number is to be multiplied by itself

Index

5^3

Base

So $5^3 = 5 \times 5 \times 5$

Indices – the plural of index

Learn 1　Powers and roots

Examples:　Find:　**a** 4^2　**b** 5^3　**c** $\sqrt{16}$　**d** $\sqrt[3]{125}$

a $4^2 = 4 \times 4 = 16$

16 is a square number because $4 \times 4 = 16$.

4 squared　　　-4 squared is $-4 \times -4 = 16$

A square number is a number 'to the power of 2' so 4 squared is also 4 to the power 2 which is written as 4^2.

The square button on a calculator looks like $\boxed{x^2}$

b $5^3 = 5 \times 5 \times 5 = 125$

125 is a cube number because $5 \times 5 \times 5 = 125$.

5 cubed　　　-5 cubed is $-5 \times -5 \times -5 = -125$

A cube number is a number 'to the power of 3' so 5 cubed is also 5 to the power 3 which is written as 5^3.

The cube button on a calculator looks like $\boxed{x^3}$

c $\sqrt{16} = 4$

Four is a square root of 16 as $4 \times 4 = 16$.

The square root of 16 is written as $\sqrt{16}$ or $\sqrt[2]{16}$, so $\sqrt{16} = 4$.

The square root button on a calculator looks like $\boxed{\sqrt{\ }}$

d $\sqrt[3]{125} = 5$

The cube root of 125 is 5 as $5 \times 5 \times 5 = 125$.

The cube root of 125 is written as $\sqrt[3]{125}$, so $\sqrt[3]{125} = 5$.

The cube root button on a calculator looks like $\boxed{\sqrt[3]{\ }}$

Apply 1

 1   16   27   −7   1   0.2   5   6

From the numbers above, write:

a a square number   **c** the square root of 49

b a cube number   **d** the cube root of 1.

 2 Write the value of each of these.

a 4^2 **e** 10^3 **i** $\sqrt[3]{1}$

b 12^2 **f** $\sqrt{9}$ **j** $\sqrt[3]{64}$

c 1.5^2 **g** $\sqrt{100}$ **k** $\sqrt[3]{-64}$

d 6^3 **h** $\sqrt{225}$

 3 Calculate these.

a $3^2 + 4^2$ **c** $10^3 - \sqrt{100}$ **e** $\sqrt{5^2 + 12^2}$

b $2^3 \times 3^2$ **d** $\sqrt{225} - \sqrt[3]{125}$ **f** $\sqrt{3^2 \times 5^2}$

 4 Which is larger?

a 2^3 or 3^2 **b** $\sqrt{64}$ or $\sqrt[3]{125}$

5 Write the value of each of these.

a 2.2^2 **e** 10.1^3 **i** $\sqrt[3]{-0.5}$

b 10.1^2 **f** $\sqrt[3]{1.5}$ **j** $\sqrt{8}$

c 8.5^2 **g** $\sqrt{75}$

d 2.2^3 **h** $\sqrt[3]{0.5}$

6 Neil says -3^2 is 9.
Andrea says -3^2 is −9.
Who is correct?
Give a reason for your answer.

7 Get Real!
A builder is laying a square concrete patio.
The area of the patio is to be 30 cm^2.
Use your calculator to find the length of the sides
correct to one decimal place.

Explore

- ⊚ The number 64 is both a square number and a cube number
- ⊚ Can you find any other numbers that are both square numbers and cube numbers?

Investigate further

Explore

- ⊚ Jenny investigates the sum of the cubes of the first two integers
  She notices that the sum gives a square number:
  $$1^3 + 2^3 = 9 \ (= 3^2)$$

- ⊚ Jenny now investigates the sum of the cubes of the first three integers
  She notices, again, that the sum gives a square number:
  $$1^3 + 2^3 + 3^3 = 36 \ (= 6^2)$$

- ⊚ Investigate the sum of the cubes of the first four integers

Investigate further

Learn 2   Rules of indices

Examples:

a Work out 5^3.

Index (or power or exponent)

5^3

Base

The index (or power or exponent) tells you how many times the base number is to be multiplied by itself.

So $5^3 = 5 \times 5 \times 5 = 125$

You can use the $\boxed{x^y}$ button for indices on your calculator

b Simplify **i** $a^2 \times a^3$ **ii** $a^3 \times a^5$ **iii** $a^6 \div a^4$ **iv** $a^7 \div a^3$
v $(a^4)^2$ **vi** $(a^2)^3$

i $a^2 \times a^3 = (a \times a) \times (a \times a \times a)$
$= a \times a \times a \times a \times a$
$= a^5$

So $a^2 \times a^3 = a^5$
and $a^3 \times a^5 = a^8$

You should notice that

$a^2 \times a^3 = a^{2+3} = a^5$
and $a^3 \times a^5 = a^{3+5} = a^8$

ii $a^3 \times a^5 = (a \times a \times a) \times (a \times a \times a \times a \times a)$
$= a \times a \times a \times a \times a \times a \times a \times a$
$= a^8$

iii $a^6 \div a^4 = \dfrac{a^6}{a^4}$

$$= \dfrac{a \times a \times a \times a \times a \times a}{a \times a \times a \times a}$$

$$= \dfrac{\cancel{a}_1 \times \cancel{a}_1 \times \cancel{a}_1 \times \cancel{a}_1 \times a \times a}{\cancel{a}_1 \times \cancel{a}_1 \times \cancel{a}_1 \times \cancel{a}_1}$$

$$= a \times a$$

$$= a^2$$

So $\qquad a^6 \div a^4 = a^2$
and $\qquad a^7 \div a^3 = a^4$

You should notice that

$\qquad a^6 \div a^4 = a^{6-4} = a^2$
and $\qquad a^7 \div a^3 = a^{7-3} = a^4$

iv $a^7 \div a^3 = \dfrac{a^7}{a^3}$

$$= \dfrac{a \times a \times a \times a \times a \times a \times a}{a \times a \times a}$$

$$= \dfrac{\cancel{a}_1 \times \cancel{a}_1 \times \cancel{a}_1 \times a \times a \times a \times a}{\cancel{a}_1 \times \cancel{a}_1 \times \cancel{a}_1}$$

$$= a \times a \times a \times a$$

$$= a^4$$

v $(a^4)^2 = a^4 \times a^4$

$$= (a \times a \times a \times a) \times (a \times a \times a \times a)$$

$$= a \times a \times a \times a \times a \times a \times a \times a$$

$$= a^8$$

So $\qquad (a^4)^2 = a^8$
and $\qquad (a^2)^3 = a^6$

You should notice that

$\qquad (a^4)^2 = a^{4 \times 2} = a^8$
and $\qquad (a^2)^3 = a^{2 \times 3} = a^6$

vi $(a^2)^3 = a^2 \times a^2 \times a^2$

$$= (a \times a) \times (a \times a) \times (a \times a)$$

$$= a \times a \times a \times a \times a \times a$$

$$= a^6$$

Apply 2

1 Write the following numbers in index notation.

a $5 \times 5 \times 5 \times 5$

b $2 \times 2 \times 2 \times 2 \times 2 \times 2 \times 2$

c $6 \times 6 \times 6 \times 6 \times 6 \times 6 \times 6 \times 6 \times 6 \times 6 \times 6 \times 6$

d 13×13

e $2 \times 2 \times 2 \times 2 \times 2 \times 2 \times 2 \times 2 \times 2 \times 2 \times 2 \times 2 \times 2 \times 2 \times 2 \times 2 \times 2 \times 2$

f $12 \times 12 \times 12 \times 12$

g $8 \times 8 \times 8 \times 8 \times 8$

h $1 \times 1 \times 1 \times 1 \times 1 \times 1 \times 1 \times 1 \times 1 \times 1 \times 1 \times 1 \times 1 \times 1 \times 1 \times 1 \times 1 \times 1 \times 1 \times 1$

2 Find the value of each of the following.

a 7^2 **e** 2^3 **i** 3^4 **m** 5^1

b 4^2 **f** 2^4 **j** 4^3 **n** 4^6

c 11^2 **g** 1^5 **k** $(-2)^6$

d $(-3)^2$ **h** 2^5 **l** $(-2)^7$

3 Simplify the following numbers, giving your answers in index form.

a $5^6 \times 5^2$ **e** $4^3 \times 4^8$ **i** $11^{10} \div 11^{14}$ **m** $(6^2)^5$

b $12^8 \times 12^3$ **f** $10^6 \times 10^{12}$ **j** $\dfrac{4^7}{4^3}$ **n** $(11^5)^4$

c $3^5 \div 3^2$ **g** $7^{11} \div 7^6$ **k** $\dfrac{9^{12}}{9^{11}}$ **o** $(10^{10})^{10}$

d $5^6 \times 5^2$ **h** $6^5 \div 6^3$ **l** $\dfrac{25^7}{25^8}$

4 Alison is investigating the rules of indices.
She writes down $4^3 \div 4^3 = 4^{3-3} = 4^0$ $4^3 \div 4^3 = 64 \div 64 = 1$ so $4^0 = 1$
She then writes down $2^5 \div 2^5 = 2^{5-5} = 2^0$ $2^5 \div 2^5 = 32 \div 32 = ...$
$3^4 \div 3^4 = 3^{4-4} = 3^0$ $3^4 \div 3^4 = ...$
$5^2 \div 5^2 = 5^{2-2} = 5^0$ $5^2 \div 5^2 = ...$

Copy and complete Alison's working. What do you notice?
Does this always work?

 5 Find the values of each of the following.

a 2^6 **d** 8^6 **g** $2^6 + 6^2$

b 2^{10} **e** 9^4 **h** $5^5 \times 10^{-4}$

c 3^5 **f** $2^{11} - 5^3$ **i** $10^8 - 10^6$

6 Say whether these statements are true or false? Give a reason for your answer.

a $6^2 = 12$ **c** $\dfrac{2^{10}}{4^5} = 1$ **e** $10^{50} \times 10^{50} = 10^{100}$

b $1^3 = 1$ **d** $3^4 + 3^5 = 3^9$

7 Simplify these leaving your answers in index form.

a $x^6 \times x^2$ **c** $\dfrac{a^7}{a^3}$ **e** $q^7 \div q^{10}$

b $e^8 \times e^3$ **d** $p^{10} \div p^5$ **f** $(b^2)^5$

8 The number 64 can be written in index form as 8^2.
Write down three other ways that 64 can be written in index form.

Explore

◎   Manjula says that 1^n is always 1
    Is Manjula correct?

◎   Try different values of n, for example, positive and negative

Investigate further

Explore

◎ One grain of rice is placed on the first square of a chessboard

◎ Two grains of rice are placed on the second square of a chessboard

◎ Four grains of rice are placed on the third square of a chessboard

◎ Eight grains of rice are placed on the fourth square of a chessboard etc

How many grains of rice will there be on the fifth square?

How many grains of rice will there be altogether on the first five squares?

How many grains of rice will there be on the tenth square?

How many grains of rice will there be altogether on the first ten squares?

Investigate further

Indices

ASSESS

The following exercise tests your understanding of this chapter, with the questions appearing in order of increasing difficulty.

1 Work these out without using a calculator.

    a 4^2 **d** $\sqrt{36}$ **g** 0^2

    b 11^2 **e** $\sqrt{196}$ **h** $(-3)^2$

    c 2.5^2 **f** $\sqrt{1.44}$

2 a Work these out without using a calculator.

    i 5^3 **iii** $\sqrt[3]{27}$ **v** $\sqrt[3]{0}$

    ii 10^3 **iv** $\sqrt[3]{64}$ **vi** $\sqrt[3]{(-8)}$

    b Work these out using the appropriate keys on a calculator.

    i 1.2^3 **iii** $\sqrt[3]{39.304}$ **v** $\sqrt[3]{-10.648}$

    ii 2.5^3 **iv** $\sqrt[3]{166.375}$

3 a Sam says numbers have two square roots.
George says some numbers have no square roots.
Who is right? Give a reason for your answer.

    b Amelia joins in the conversation and says that all numbers have two cube roots.
Is she right? Give a reason for your answer.

4 a Work out the following, giving your answers in index form.

    i $4^6 \times 4^2$ **v** $6^4 \times 6^2 \times 6^3$ **ix** $5^8 \div 5^7$

    ii $11^5 \times 11^3$ **vi** $10^4 \div 10^2$ **x** $2^3 \div 2^3$

    iii $(5^3)^2$ **vii** $21^7 \div 21^5$

    iv $7^5 \times 7$ **viii** $16^{10} \div 16^9$

b Find the value of:

    i $3^2 \times 4^2$ **iii** $6^5 \times 6^3 \div 6^4$

    ii $3^4 \div 5^2$ **iv** $\dfrac{(10^8 \times 10^7)}{(10^7 \times 10^6)}$

c Which is greater:

    i 3^5 or 5^3 **ii** 11^2 or 2^{11} **iii** 2^4 or 4^2?

5 Simplify:

    a $x^4 \times x^3$ **c** $z \times z^7$ **e** $q^8 \div q^6$

    b $y^5 \times y^2$ **d** $p^5 \div p^4$ **f** $(t^4)^2$

6 Tom says that $k^4 \div k^6 = k^2$.
Is Tom correct? Explain your answer.

Try a real past exam question to test your knowledge:

7 a Work out the cube of 4.

    b Work out 0.2^2

    c A list of numbers is given below.

            15   16   19   27   34   42   45

    From this list, write:

    i a cube number

    ii a prime number.

Spec B, Int Paper 2, Nov 02

14 Sequences

OBJECTIVES

G **Examiners would normally expect students who get a G grade to be able to:**

Continue a sequence of numbers or diagrams

Write the terms of a simple sequence

F **Examiners would normally expect students who get an F grade also to be able to:**

Find a particular term in a sequence involving positive numbers

Write the term-to-term rule in a sequence involving positive numbers

E **Examiners would normally expect students who get an E grade also to be able to:**

Find a particular term in a sequence involving negative or fractional numbers

Write the term-to-term rule in a sequence involving negative or fractional numbers

D **Examiners would normally expect students who get a D grade also to be able to:**

Write the terms of a sequence or a series of diagrams given the nth term

C **Examiners would normally expect students who get a C grade also to be able to:**

Write the nth term of a sequence or a series of diagrams

What you should already know ...

■ Identify odd and even numbers

VOCABULARY

Sequence – a list of numbers or diagrams that are connected in some way

In this sequence of diagrams, the number of squares is increased by one each time:

The dots are included to show that the sequence continues

Term – a number, variable or the product of a number and variable(s) such as 3, x or $3x$

nth term – this phrase is often used to describe a 'general' term in a sequence; if you are given the nth term, you can use this to find the terms of a sequence

Square numbers – 1, 4, 9, 16, 25, ...

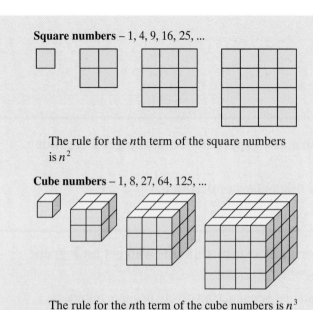

The rule for the nth term of the square numbers is n^2

Cube numbers – 1, 8, 27, 64, 125, ...

The rule for the nth term of the cube numbers is n^3

Triangle numbers – 1, 3, 6, 10, 15, ...

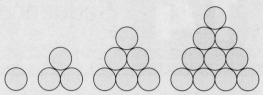

The rule for the nth term of the triangle numbers is $\frac{1}{2}n(n + 1)$

Fibonacci numbers – a sequence where each term is found by adding together the two previous terms

1,　　1,　　2,　　3,　　5,　　8,　　13,　　21,　　...

$1+1$　$1+2$　$2+3$　$3+5$　$5+8$　$8+13$

Learn 1　The rules of a sequence

Example:

Write the next two terms in the sequence:

5, 10, 20, 40, ...

In this sequence of numbers, the number is multiplied by two to give the next number.

The numbers of the sequence are usually called the terms of the sequence

5,　　10,　　20,　　40,　　...

$×2$　$×2$　$×2$　$×2$

In the sequence above, the number 5 is the first term, the number 10 is the second term, the number 20 is the third term, the number 40 is the fourth term, etc

The rule for the sequence (called the term-to-term rule) can be used to find the fifth and subsequent terms.

In this example, the term-to-term rule is 'multiply by 2' so the fifth term is 80 ($40 × 2$) and the sixth term is 160 ($80 × 2$) and so on.

5,　　10,　　20,　　40,　　80,　　160,　　...

$×2$　$×2$　$×2$　$×2$　$×2$　$×2$

Apply 1

1 Draw the next two diagrams in the following sequences.

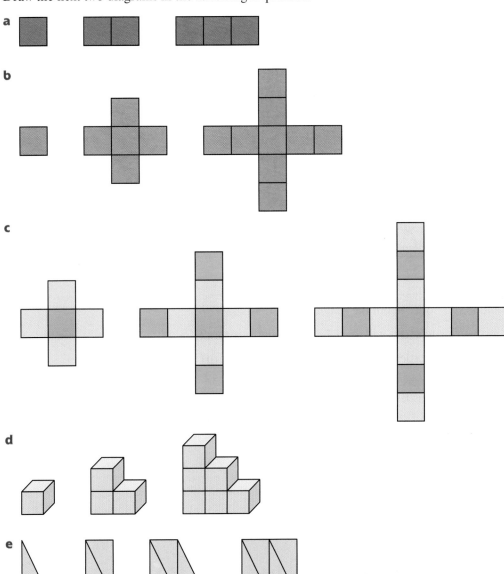

2 Write the next two terms in the following sequences.

    a 3, 7, 11, 15, ...

    b 6, 9, 12, 15, ...

    c 2, 10, 18, 26, ...

    d 0, 5, 10, 15, ...

    e 1, 2, 4, 8, 16, ...

    f 10, 100, 1000, 10 000, ...

3 Write the ninth and tenth terms in the following sequences.

   a 2, 4, 6, 8, ...

   b 3, 5, 7, 9, ...

   c 10, 20, 30, 40, ...

   d 101, 102, 103, 104, ...

4 Copy and complete this table.

Pattern (n)	Diagram	Number of matchsticks (m)
1	△	3 matchsticks
2	◁▷	5 matchsticks
3	◁▷◁	7 matchsticks
4	◁▷◁▷	
5		

   a What do you notice about the pattern of matchsticks?

   b How many matchsticks will there be in the ninth pattern?

   c How many matchsticks will there be in the tenth pattern?

   d There are 41 matchsticks in the 20th pattern.
     How many matchsticks are there in the 21st pattern?
     Give a reason for your answer.

5 Fill in the missing numbers in the following sequences.

   a 4, 6, 8, ..., 12, 14

   b 3, 6, 12, ..., 48, 96

   c 25, 16, 7, ..., −11

   d 2, ..., 20, 29, 38

6 Write the term-to-term rule for the following sequences.

   a 3, 7, 11, 15, ...   **d** 3, 4.5, 6, 7.5, ...   **g** 20, 16, 12, 8, ...

   b 0, 5, 10, 15, ...   **e** 2, 3, 4.5, 6.75, ...   **h** 54, 18, 6, 2, ...

   c 1, 2, 4, 8, 16, ...   **f** 100, 1000, 10 000, ...

7 The term-to-term rule is +6.
   Write five different sequences that fit this rule.

8 Here is a sequence of numbers:

   2   3   5   9   17

The rule for continuing this sequence is: **multiply by 2 and subtract 1**.

a What are the next **two** numbers in this sequence?

The same rule is used for a sequence that starts with the number -3.

b What are the next **two** numbers in this sequence?

9 Farukh is exploring number patterns.
He writes down the following products in a table.

1×1	1
11×11	121
111×111	12 321
1111×1111	1 234 321
11111×11111	
111111×111111	

Copy and complete the last two rows in the table.

Farukh says he can use the table to work out 1 111 111 111 × 1 111 111 111
Is he correct? Give a reason for your answer.

Explore

◉ Investigate the following non-linear sequences:
  1, 4, 9, 16, ...
  1, 8, 27, 64, ...
  1, 3, 6, 10, 15, ...
  1, 1, 2, 3, 5, 8, ...
  2, 3, 5, 7, 11, ...

◉ Invent your own non-linear sequences

Investigate further

Explore

◉ Workers are paid 1p on their first day, 2p on the second day, 4p on the third day, 8p on the fourth day and so on

◉ How much will a worker get paid on the fifth day?

◉ How much money will a worker get for the first five days altogether?

◉ How much money will the worker get for the first ten days altogether?

Investigate further

Learn 2 The *n*th term of a sequence

Example: Given the *n*th term of a sequence is $2n + 3$, use this to find the first four terms of the sequence.

If the *n*th term is $2n + 3$, you can use this to find the first term by replacing *n* by 1 in the formula.
Similarly, you can find the second term by replacing *n* by 2 in the formula and the 100th term by replacing *n* by 100 in the formula.
So the sequence whose *n*th term is $2n + 3$ is:

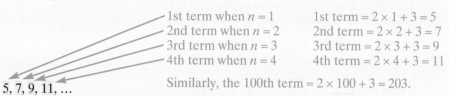

1st term when $n = 1$ 1st term $= 2 \times 1 + 3 = 5$
2nd term when $n = 2$ 2nd term $= 2 \times 2 + 3 = 7$
3rd term when $n = 3$ 3rd term $= 2 \times 3 + 3 = 9$
4th term when $n = 4$ 4th term $= 2 \times 4 + 3 = 11$

5, 7, 9, 11, …   Similarly, the 100th term $= 2 \times 100 + 3 = 203$.

The above sequence is called a linear sequence because the difference between terms are all the same. In this example, the differences are all +2. We say that the term-to-term rule is +2.

5,   7,   9,   11,   …
  +2   +2   +2

In general, to find the *n*th term of a linear sequence, you can use the formula:

nth term = differences $\times n$ + (first term – differences)

$$= dn + (a - d)$$

a is the first term $= 5$
d is the difference between terms $= 2$

So for   5,   7,   9,   11,   …
         +2   +2   +2

nth term = difference $\times n$ + (first term – difference)

$$= 2 \times n + (5 - 2)$$

$$= 2n + 3$$

Apply 2

Apart from question 9, this is a non-calculator exercise.

1 Write the first five terms of the sequence whose *n*th term is:

  a $n + 3$ **d** $2n - 5$ **g** $n^2 + 3$

  b $5n$ **e** $3n^2$ **h** $\dfrac{n}{n + 2}$

  c $5n - 3$ **f** $n^2 - 5$

2 Write the 100th and the 101st terms of the sequence whose *n*th term is:

  a $n + 5$ **c** $100 - 2n$

  b $3n - 7$ **d** $n^2 + 1$

3 Jenny writes the sequence 3, 7, 11, 15, ...
She says that the *n*th term is *n* + 4.
Is she correct?
Give a reason for your answer.

4 Write the *n*th term in these linear sequences.

    a 3, 7, 11, 15, ...   **d** −5, −1, 3, 7, ...   **g** 4, 6.5, 9, 11.5, ...

    b 0, 5, 10, 15, ...   **e** 100, 95, 90, ...   **h** −5, 3, 11, 19, ...

    c 8, 14, 20, 26, ...   **f** 23, 21, 19, 17, ...

5 Write the formula for the number of squares in the *n*th pattern.

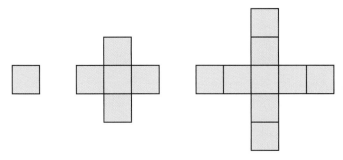

6

Pattern (*n*)	Diagram	Number of matchsticks (*m*)
1	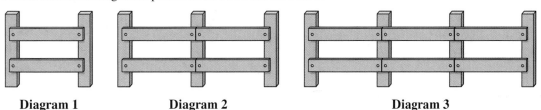	3 matchsticks
2		5 matchsticks
3		7 matchsticks

Write the formula for the number of matchsticks (*m*) in the *n*th pattern.

7 Get Real!
Jackie builds fencing from pieces of wood as shown below.

    Diagram 1   **Diagram 2**   **Diagram 3**
  4 pieces of wood   7 pieces of wood   10 pieces of wood

How many pieces of wood will there be in Diagram *n*?

8

  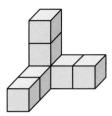

Stuart says that the number of cubes in the 100th pattern is 300.
How can you tell that Stuart is wrong?
Give a reason for your answer.

9 Write the nth term in these non-linear sequences.

    a 1, 4, 9, 16, ...   **d** 1, 8, 27, 64, 125, ...

    b 2, 5, 10, 17, ...   **e** 0, 7, 26, 63, 124, ..

    c 2, 8, 18, 32, ...   **f** 10, 100, 1000, ...

10 Write the nth term in the following sequences.

    a $1 \times 2, 2 \times 3, 3 \times 4, ...$

    b $\frac{2}{3}, \frac{3}{4}, \frac{4}{5}, \frac{5}{6}, ...$

    c $1 \times 2 \times 5, 2 \times 3 \times 6, 3 \times 4 \times 7, 4 \times 5 \times 8, ...$

    d 0.1, 0.2, 0.3, 0.4, ...

Explore

 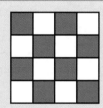

  ◉ Write the formula for the number of white tiles in the nth pattern

  ◉ Write the formula for the number of red tiles in the nth pattern

Investigate further

Sequences

ASSESS

The following exercise tests your understanding of this chapter, with the questions appearing in order of increasing difficulty.

1 a Draw the next two diagrams in the following sequences.

b Write the next two terms in the following sequences.

   i Sunday, Tuesday, Thursday, ...

   ii A, E, I, M, ...

   iii 2, 7, 12, 17, ...

   iv 1.2, 1.4, 1.6, 1.8, ...

   v 2, 6, 18, 54, ...

   vi 3, 30, 300, 3000, ...

2 a Write the 6th and 10th terms in the following sequences.

   i 2, 5, 8, 11, ...

   ii 1, 6, 11, 16, ...

   iii 3, 6, 12, 24, ...

b Write the term-to-term rule for the following sequences.

   i 1, 5, 9, 13, ...

   ii 2, 10, 50, 250, ...

   iii 3, 8, 13, 18, 23, ...

3 a Write the 6th and 10th terms in the following sequences.

   i 20, 17, 14, 11, ...

   ii 64, 32, 16, ...

   iii 1, -2, -5, -8, ...

b Write the term-to-term rule for the following sequences.

   i 8, 5, 2, -1, ...

   ii -1, -4, -7, -10, ...

   iii 6, 3, 1.5, 0.75, 0.375, ...

4 Write the first three terms and the 5th, 20th and 50th terms of the sequences with nth term.

    a $2n + 1$ **b** $5n - 2$ **c** $n^2 + 1$.

5 Find the nth term of the following sequences.

    a 6, 8, 10, 12, ...   **c** 8, 6, 4, 2, ...   **e** 4, 7, 12, 19, ...

    b 3, 13, 23, 33, ...   **d** −2, 5, 12, 19, ...

Try a real past exam question to test your knowledge:

6 The nth term of a sequence is $3n - 1$.

    a Write down the first and second terms of the sequence.

    b Which term of the sequence is equal to 32?

    c Explain why 85 is not a term in this sequence.

Spec A, Int Paper 2, Nov 04

15 Coordinates

What you should already know ...

■ Negative numbers and the number line

VOCABULARY

Coordinates – a system used to identify a point; an x-coordinate and a y-coordinate give the horizontal and vertical positions

Origin – the point (0, 0) on a coordinate grid

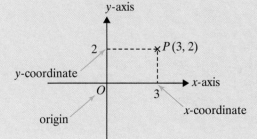

Axis (pl. axes) – the lines used to locate a point in the coordinates system; in two dimensions, the *x*-axis is horizontal, and the *y*-axis is vertical. This system of Cartesian coordinates was devised by the French mathematician and philosopher René Descartes

In three dimensions, the *x*- and *y*-axes are horizontal and at right angles to each other and the *z*-axis is vertical

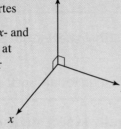

Horizontal – from left to right; parallel to the horizon

Horizontal

Vertical – directly up and down; perpendicular to the horizontal

Vertical

Gradient – a measure of how steep a line is

$$\text{Gradient} = \frac{\text{change in vertical distance}}{\text{change in horizontal distance}}$$
$$= \frac{y}{x}$$

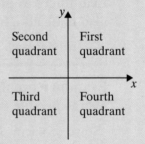

Intercept – the *y*-coordinate of the point at which the line crosses the *y*-axis

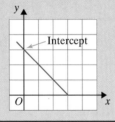

Equidistant – the same distance; if A is equidistant from B and C, then AB and AC are the same length

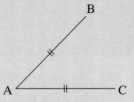

Line segment – the part of a line joining two points

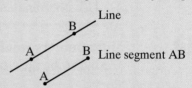

Midpoint – the middle point of a line

Quadrant – one of the four regions formed by the *x*- and *y*-axes in the Cartesian coordinate system

Second quadrant	First quadrant
Third quadrant	Fourth quadrant

Learn 1 Coordinates in four quadrants

Example: Name the point drawn on the diagram.

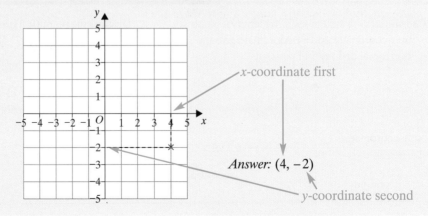

x-coordinate first

Answer: (4, −2)

y-coordinate second

Apply 1

1 Write down the coordinates of points A, B, C, D and E, which are marked on the grid on the right.

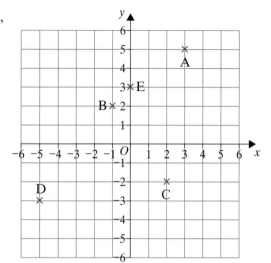

2 Draw a grid like the one in question **1**, with the *x*-axis and *y*-axis labelled from −6 to 6.

  a On your grid, mark the points A(3, 2), B(4, −1), C(1, −2) and D(0, 1).

  b Join A to B, B to C, C to D and D back to A.

  c What is the name of the shape you have drawn?

3 A plot of land 10 metres square contains 25 anthills that have to be cleared.

The anthills are located at

(1.5, 0.5)	(4.5, 0)	(3.5, 1.5)	(6, 1.5)	(7.5, 1.5)
(9, 2)	(4.5, 2.5)	(1, 3)	(3.5, 3.5)	(9.5, 4)
(1, 4.5)	(7, 4.5)	(2.5, 6)	(4, 6)	(5, 6)
(6.5, 6)	(1, 6.5)	(10, 7)	(0, 8)	(3, 8)
(5, 8)	(8.5, 8.5)	(2, 9)	(5, 9)	(3.5, 10)

Draw axes from 0 to 10 and mark the positions of the anthills with a 'O'.

Starting at (0, 0), plot a path through the anthills from bottom to top that would keep you at least 1 metre from any anthill.

Write down the coordinates of each point you move to in a straight line from the previous point.

4 a Pip says that the two points, A and B, marked on the grid on the right are (1, 1) and (−3, 1). Is she right?

  b A and B are two corners of a square. Write down the coordinates of two other points that could be the other two corners of the square. (There are three possible answers to this question. Can you find them all?)

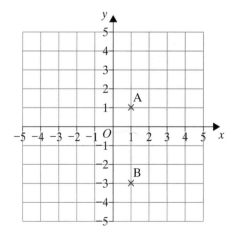

5 Holly says it is further from $(4, -2)$ to $(3, 5)$ than it is from $(4, 2)$ to $(-3, 5)$.
Lisa says it is further from $(4, 2)$ to $(-3, 5)$ than it is from $(4, -2)$ to $(3, 5)$.
Farid says the distances are the same.
Plot the points on a grid and measure the distances to find out who is correct.

6 Write down the coordinates of

    a the bottom left-hand corner of the house

    b the top right-hand corner of the chimney

    c all four corners of the door.

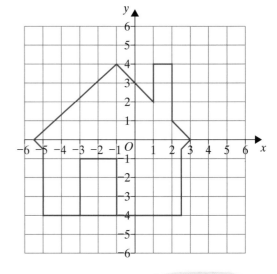

7 Get Real!

A boat receives a call for help from a yacht.

    a Draw a grid with the x-axis and y-axis labelled from -6 to 6.
       Use a scale of 1 cm to represent 1 km.

    b The boat is at the point $(0, 0)$. Mark the position of the boat.

    c The yacht is at the position $(3, -4)$. Mark the position of the yacht.

    d The boat sails to the yacht. What is the bearing and the distance
       from the boat to the yacht?

Explore

  ◎ Copy the grid and mark the points $(1, 3)$, $(1, 1)$, and $(3, 1)$

  ◎ You are not allowed to mark a point that would complete
     the four corners of a square
     For example, you must not mark the point $(3, 3)$

  ◎ Mark any other point and write down its coordinates

  ◎ Carry on marking points and writing down their
     coordinates, making sure you don't mark the four corners
     of a square of any size

  ◎ What is the greatest number of points you can mark?

Investigate further

Learn 2   Equations and straight lines

Example: Draw the line that has an equation of $y = 3x - 4$

A straight line is named by an equation which fits all pairs of coordinates on the line. For example, the line $y = 3x - 4$ joins all the points whose y-coordinate is four less than three times the x-coordinate ($3 \times x - 4$).

Choose three x-coordinates, such as $(0, \ )$, $(1, \ )$ and $(3, \ )$.
Calculate the y-coordinates: $x = 0, y = 3 \times 0 - 4 = -4$ The point is $(0, -4)$.
$x = 1, y = 3 \times 1 - 4 = -1$ The point is $(1, -1)$.
$x = 3, y = 3 \times 3 - 4 = 5$ The point is $(3, 5)$.

Plot the points and join them up with a straight line using a ruler.

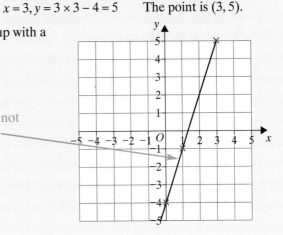

If your three points are not in a straight line, check your working

Apply 2

1 a Draw a grid with the x-axis and y-axis labelled from -6 to 6.

b Mark four points with an x-coordinate of 3, for example $(3, -2)$.
Draw a straight line through all four points.

c What is the equation of the line you have drawn?

d Now draw four points with a y-coordinate of -2. Draw a straight line through all four points.

e What is the equation of the line you have drawn?

f Where do the two straight lines cross?

2 a Draw a grid with the x-axis and y-axis labelled from -6 to 6.
Draw the lines $x = 2$ and $y = 4$.
Write down the coordinates of the points where the lines cross each other.

Now write down where these lines cross each other. You may be able to work them out without drawing the lines.

b $x = 5$ and $y = -4$ **d** $x = 0$ and $y = 1$ **f** $x = 2$ and $y = x$

c $x = -3$ and $y = -2$ **e** $x = 2.5$ and $y = -3.5$ **g** $x = 4$ and $y = x - 1$

3 The equation $y = x + 2$ means that y is 2 more than x. For example, the point (1, 3) has a y-coordinate which is 2 more than the x-coordinate.

   a Write down five more points that fit the equation, that is, write down five more pairs of coordinates where the y-coordinate is two more than the x-coordinate.

   b Plot these points on a grid.

   c Draw a straight line through the points.

   d Repeat steps **a**, **b** and **c** for the equation $y = x - 1$

4 Toby says the line $y = x + 3$ passes through the point (4, 1).
Colin says the line $y = x + 3$ passes through the point (1, 4).
Who is right, Toby or Colin?

5 Javindra says that the line $y = x + 3$ is the same line as $x = y - 3$.
Is she right?
Give a reason for your answer.

<u>**6**</u> Where do these pairs of lines cross?

   a $x = 4$ and $y = x + 2$ **d** $x = -2$ and $y = 3x - 1$

   b $x = -2$ and $y = x - 1$ **e** $y = 4$ and $y = x + 3$

   c $x = 3$ and $y = 2x$ **f** $y = -2$ and $y = 3x + 1$

7 For each grid below

   i write down the coordinates of three points on the line

   ii use your answers to part **i** to help you write down the equation of the line.

a

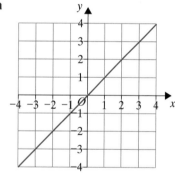

c

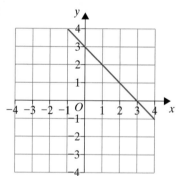

b

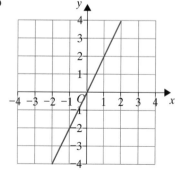

8 For each grid below

  i write down the coordinates of three points on the line

  ii use your answers to part **i** to help you write down the equation of the line.

a
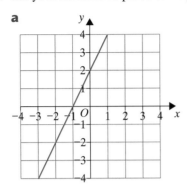

b
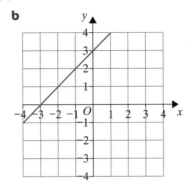

c
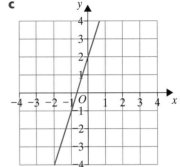

9 Get Real!

A farmer owns a field, and wants to put a fence around it.
He describes the shape as follows:

'It's a four-sided field, but it isn't a rectangle.
You can draw a plan of it like this:
Draw a coordinate grid with the x-axis and y-axis labelled from -5 to $+5$.
Draw in the four lines $x = 4$, $y = 5$, $y = 2x + 3$, and $x + y = -2$.'

Draw a plan of the farmer's field.

Explore

  ◎ Draw a coordinate grid with the x-axis and y-axis labelled from -6 to 6

  ◎ Draw a flag by joining $(1, 3)$ to $(1, 5)$ to $(2, 4)$ to $(1, 4)$

  ◎ Double the x-coordinates and draw it again; $(2, 3)$ to $(2, 5)$ to $(4, 4)$ to $(2, 4)$

  ◎ What happens to the flag?

  ◎ Try other manipulations, for example swap the coordinates over so $(1, 3)$ becomes $(3, 1)$, or subtract the y-coordinates from 8, or change the signs of the x-coordinate.

Investigate further

Explore

  ◎ You *may* want to use graph-plotting software for this

  ◎ Draw lines of equations such as $y = 2x + 1$

  ◎ What happens if you change the $+1$?

  ◎ What happens if you change the 2?

Investigate further

Learn 3   The midpoint of a line segment

The midpoint of the line segment from (a, b) to (c, d) is $\left(\dfrac{a+c}{2}, \dfrac{b+d}{2}\right)$.

The mean of the
x-coordinates

The mean of the
y-coordinates

Example: Write down the coordinates of the point halfway between A and B.

A is the point $(-4, 3)$; B is $(1, 2)$.

The midpoint is at

$$\left(\frac{-4+1}{2}, \frac{3+2}{2}\right)$$

$$= \left(\frac{-3}{2}, \frac{5}{2}\right)$$

$$= (-1.5, 2.5)$$

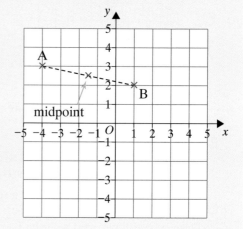

midpoint

Apply 3

1 a Write down the coordinates of the point halfway between $(1, 6)$ and $(7, 2)$.

  b Draw a grid with the x-axis and y-axis labelled from 0 to 8. Plot the points $(1, 6)$ and $(7, 2)$ and use your diagram to check your answer to part **a**.

2 A triangle ABC has vertices at A$(8, -4)$, B$(2, 6)$ and C$(-4, 2)$.

  a Find the coordinates of X, the midpoint of AB.

  b Find the coordinates of Y, the midpoint of AC.

  c Draw a grid with the x-axis and y-axis labelled from -4 to 8. Plot the points A, B, C, X and Y.

  d Draw the lines XY and BC. What do you notice about them?

3 Work out the coordinates of the point halfway between $(4, -5)$ and $(1, 3)$.

4 If A is the point $(5, -1)$ and B is the point $(-7, -4)$, what are the coordinates of the midpoint of the line AB?

5 Liam says that the point $(2, 1.5)$ is halfway between $(-4, 2)$ and $(8, -5)$.
Is he correct?
Give a reason for your answer.

6 C is the mid-point of the line AB.
The coordinates of C are $(4, -1)$. B is the point $(2, 5)$.
What are the coordinates of A?

7 A quadrilateral ABCD has coordinates as follows:
$A(4, -2)$; $B(7, -4)$; $C(-2, 4)$; $D(-5, 6)$.

a Find the midpoint of the diagonal AC.

b Find the midpoint of the diagonal BD.

c What can you say about the quadrilateral ABCD?

8 a $(6, -8)$ is the midpoint of a line segment AB. A is the point $(3, -6)$.
Find the coordinates of B.

b Write five other sets of coordinates that have a midpoint at $(6, -8)$ on
line segment AB.

9 Get Real!
A computer programmer uses coordinates to plot points in the screen
when designing a computer game.

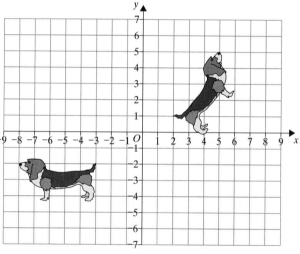

She draws a picture of a dog, with its nose at $(-8, -2)$ and its tail at $(-3, -2)$.
She draws a reflection of the dog with its nose at $(5, 5)$ and its tail at $(2, 1)$.
She wants to draw the mirror for the reflection. She knows the mirror must
go halfway between the dog and its reflection.

a What are the coordinates of the point halfway between the dog's nose
and its reflection?

b What are the coordinates of the point halfway between the dog's tail
and its reflection?

10 Lori draws a circle with its centre at $(-3, -2)$. She draws a diameter from
the point $(2, -8)$. Find the coordinates of the other end of the diameter.

11 Find your way through the maze by only occupying squares where M is the midpoint of AB.

1	2	3	4	5	6
START	A (3, 2) B (3, 7) M (6, 4.5)	A (−2, 11) B (4, −4) M (−1, 3.5)	A (6, −9) B (3, 9) M (4.5, 9)	A (−11, 29) B (−15, 22) M (−13, 25.5)	**END**
7 A (4, 5) B (2, 9) M (3, 7)	8 A (1, 2) B (2, 3) M (3, 5)	9 A (4, 2) B (−4, −2) M (0, 2)	10 A (−1.4, 0.2) B (1.3, 0.6) M (−0.05, 0.4)	11 A (2.3, −0.8) B (−1.3, 0.3) M (0.5, −0.25)	12 A (3, 1) B (4, 2) M (5, 3)
13 A (4, −4) B (−2, 6) M (1, 1)	14 A (8, −2) B (3, −4) M (5.5, −3)	15 A (−13, 24) B (−12, 9) M (−12.5, 17)	16 A (0, 7) B (−7, 0) M (−3.5, 3.5)	17 A (−1, 9) B (9, −1) M (4, −4)	18 A (−12, 19) B (−12, 2) M (0, 10.5)
19 A (0, 0) B (2, 3) M (2, 3)	20 A (−4, −4) B (−2, 7) M (−3, 1.5)	21 A (4, −11) B (−4, −11) M (0, 0)	22 A (−8.5, 3.4) B (4.1, 2.2) M (−2.2, 2.8)	23 A (23, −13) B (13, −12) M (18, −12.5)	24 A (4.2, −3.7) B (−2.4, 1.3) M (0.9, −1.2)
25 A (4, −12) B (−11, 8) M (−4.5, −2)	26 A (−4, 7) B (2, 5) M (−1, 6)	27 A (−3, 5) B (2, −6) M (−0.5, −0.5)	28 A (0, 9.2) B (−2.4, 3.2) M (−1.2, 6.4)	29 A (1, 5) B (3, 1) M (4, 3)	30 A (23, 17) B (−14, 11) M (4.5, 14)
31 A (4.2, 3) B (1.9, 8) M (3.1, 5.5)	32 A (−9, −8) B (−3, 9) M (6, 0.5)	33 A (−1, 4) B (−3, −4) M (−2, 0)	34 A (−3, 4) B (−2, −3) M (−2.5, 0.5)	35 A (4.2, −3.3) B (2.4, 1.3) M (3.3, −1)	36 A (−3, 2) B (2, −3) M (−0.5, −0.5)

Explore

◎ Draw a coordinate grid, labelled from −8 to 8 on both axes

◎ Plot the points A(6, 2), B(−4, 2), C(−4, −8) and D(6, −8)

◎ Join A to B, B to C, C to D and D to A

◎ Write down the name of the shape you have drawn

◎ Work out the midpoint of each side of the shape, and mark them on the grid

◎ Join these midpoints up.
What shape have you drawn?

Investigate further

Learn 4 Coordinates in three dimensions

Remember to write coordinates alphabetically: (x, y, z)

Example:

In the diagram of the cuboid
$OA = 2$ units, $AB = 5$ units and $AD = 3$ units.
O is the origin.
Write down the coordinates of A, B, C and D.

$A = (2, 0, 0)$ $C = (2, 5, 3)$
$B = (2, 5, 0)$ $D = (2, 0, 3)$

x-coordinate first, then y, then z
Draw and label diagrams to make the work easier

Apply 4

Questions 1 to 2 are about the diagram on the right.

1 In the diagram, A is the point (6, 4, 5).
Write down the coordinates of points B, C, D and E.

2 An identical cuboid is placed on top of the one in the diagram.
Write down the coordinates of the top corner directly above A.

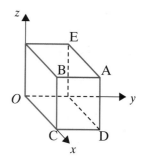

3 What is the midpoint of the line joining (6, 3, 7) to (2, 5, 9)?

4 What is the midpoint of the line joining (−2, 4, −5) to (6, 2, −6)?

5 Alex is designing a toy street. A plan of it looks like this:

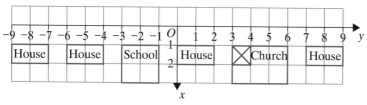

The front elevation looks like this:

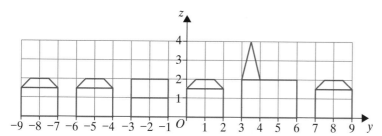

Write down the coordinates of the top of the church spire.

6 An electricity company plans where to put its electricity pylons on a map.
The results are shown below.

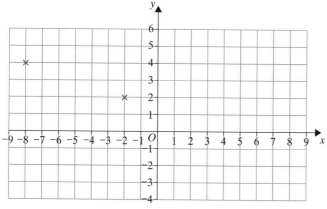

They plan to put the first one at (−8, 4), and the second at (−2, 2).

a If they space them equally and in a straight line, where will they place the third?

b The top of the first one is (−8, 4, 1). Assuming they are all the same height and on level ground, what are the three-dimensional coordinates of the top of the fourth pylon?

Explore

◎ A $2 \times 2 \times 2$ cube is placed with a corner at O, the origin, and other corners at A(2, 0, 0), B(0, 2, 0), C(2, 2, 0), D(0, 0, 2), E(0, 2, 2), F(2, 0, 2) and G(2, 2, 2)

◎ How many right angles can you find by joining up corners of the cube?

◎ How many angles of 45° can you find?

◎ How many angles of 60° can you find?

Investigate further

Coordinates

ASSESS

The following exercise tests your understanding of this chapter, with the questions appearing in order of increasing difficulty.

1 David has copied part of a map onto the square grid shown below.

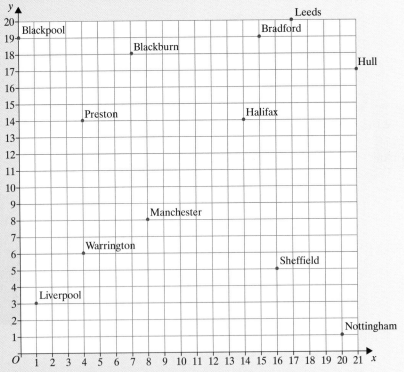

a Write down the coordinates of the 12 places shown.

b The following places also have coordinates on the grid. Copy the grid and mark in the places on it.

Chester (1, 0)   Macclesfield (9, 1)   Mansfield (17, 3)
Southport (0, 12)   Burnley (11, 19)   Barnsley (16, 9)
Clitheroe (8, 20)   Scunthorpe (20, 11)

2 Draw axes on a piece of graph paper. Take values of x from -9 to $+8$ and values of y from -8 to $+10$. Plot the following sets of points on the same grid and join the points together, resulting in a well known shape.

A: $(-9, -3)$ $(-6, -8)$ $(8, -8)$ $(8, -2)$ $(7, -3)$ $(-9, -3)$
B: $(-8, -2)$ $(7, -2)$ $(1, 10)$ $(1, -2)$
C: $(1, 7)$ $(-6, 10)$ $(-8, -2)$ $(1, -2)$

3 For each grid write the equation of the straight line.

a

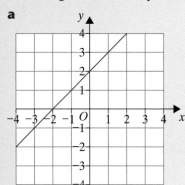

b

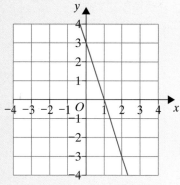

c

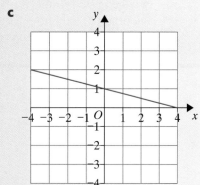

4 Draw a coordinate grid on x- and y-axes labelled from -6 to $+6$.
Draw and label these lines:

a $y = 3x + 2$

b $y = 5 - 3x$

c $y + 2x = 4$

5 Find the midpoints of the following line segments.
Draw sketch diagrams to show your answers.

a $(2, 3)$ and $(6, 7)$ **d** $(2, 9)$ and $(-2, -9)$

b $(9, -9)$ and $(3, 4)$ **e** $(-4, 10)$ and $(6, -8)$

c $(-4, 4)$ and $(6, 10)$

6 a On graph paper, plot the coordinates of the quadrilateral ABCD, given by the points $(-6, -7)$, $(-4, 8)$, $(2, 3)$ and $(5, -5)$.

b Calculate the coordinates of P, Q, R and S, the midpoints of AB, BC, CD and DA.

c Plot P, Q, R, S and draw the quadrilateral PQRS.

d What do you notice about the quadrilateral PQRS?

7 a Write down the midpoint of the line segment joining $(3, -1, 2)$ and $(4, 5, -5)$.

b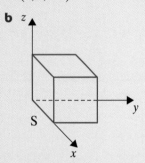

A cube of side 5 units is placed with one corner of its base at the origin of a three-dimensional grid, as shown.
Write down the coordinates of the other seven vertices.

16 Collecting data

OBJECTIVES

G **Examiners would normally expect students who get a G grade to be able to:**

Design and use tally charts for discrete and grouped data

E **Examiners would normally expect students who get an E grade also to be able to:**

Design and use two-way tables for discrete and grouped data

D **Examiners would normally expect students who get a D grade also to be able to:**

Classify and know the difference between various types of data

Use a variety of different sampling methods

Design and use data collection sheets and questionnaires

C **Examiners would normally expect students who get a C grade also to be able to:**

Identify possible sources of bias in the design and use of data collection sheets and questionnaires

---

What you should already know ...

■ Counting and, in particular, counting in fives (for tally charts)

VOCABULARY

Tally chart – a useful way to organise raw data; the chart can be used to answer questions about the data, for example,

Number of pets	Tally
0	JHT IIII
1	JHT JHT II
2	JHT II
3	III
4	II

The tallies are grouped into five so that

IIII = 4

JHT = 5

JHT I = 6

This makes the tallies easier to read

Two-way table – a combination of two sets of data presented in a table form, for example,

	Men	Women
Left-handed	7	6
Right-handed	20	17

Quantitative data – data that can be counted or measured using numbers, for example, number of pets, height, weight, temperature, age, shoe size, etc.

Qualitative or **categorical data** – data that cannot be measured using numbers, for example, type of pet, car colour, taste, peoples' opinions/feelings, etc.

Discrete data – data that can only be counted and take certain values, for example, numbers of cars (you can have 3 cars or 4 cars but nothing in between, so $3\frac{1}{2}$ cars is not possible)

Continuous data – data that can be measured and take any value; length, weight and temperature are all examples of continuous data

Survey – a way of collecting data; there are a variety of ways of doing this, including face-to-face, or via telephone, e-mail or post using questionnaires

Respondent – the person who answers the questionnaire

Direct observation – collecting data first-hand, for example, counting cars at a motorway junction or observing someone shopping

Primary data – data that you collect yourself; this is new data and is usually collected for the purpose of a task or project (including GCSE coursework)

Secondary data – data that someone else has collected; this might include data in books, newspapers, magazines, etc. or data that has been loaded onto a database

Data collection sheets – these are used to record the responses to the different questions on a questionnaire; they can also be used with computers to load data onto a database

Pilot survey – a small-scale survey to check for any unforeseen problems with the main survey

Convenience or **opportunity sampling** – a survey that is conducted using the first people who come along, or those who are convenient to sample (such as friends and family)

Random sampling – this requires each member of the population to be assigned a number; the numbers are then chosen at random

Systematic sampling – this is similar to random sampling except that it involves every nth member of the population; the number n is chosen by dividing the population size by the sample size

Quota sampling – this method involves choosing a sample with certain characteristics, for example, select 20 adult men, 20 adult women, 10 teenage girls and 10 teenage boys to conduct a survey about shopping habits

Cluster sampling – this is useful where the population is large and it is possible to split the population into smaller groups or clusters

Learn 1 Organising data

Examples:

a The following information, collected from students in a class, shows the number of pets owned by the students.

1	4	0	3	2	0	1	2	1	0	0
2	3	1	1	0	1	0	2	1	1	0
3	2	2	1	0	1	0	1	1	2	4

In this form, the information is called 'raw data' because it has still to be organised

Show this information in a tally chart.

Number of pets	Tally
0	⊮ IIII
1	⊮ ⊮ II
2	⊮ II
3	III
4	II

The tallies are grouped into five so that

IIII $= 4$
⊮ $= 5$
⊮ I $= 6$

This makes the tallies easier to read

Sometimes the tally chart is extended with another column which contains the totals or frequencies.

b The following information, collected from newspapers, shows the numbers of words in sentences.

16	22	14	12	19	23	18	21	24	29	17
22	17	11	15	18	19	20	22	15	17	18
25	16	21	20	19	15	12	14	8	11	19

For larger data sets, it may be necessary to group data. When choosing suitable groups it is best to have between four and six groups which are all the same size. When completing the tally chart always check you have the correct total.

Show this information in a tally chart.

Number of words	Tally	Frequency
5–9	I	1
10–14	IIII I	6
15–19	IIII IIII IIII	15
20–24	IIII IIII	9
25–29	II	2

c The information collected about pets on the previous page might be shown in terms of both male and female students.

Number of pets	Frequency
0	9
1	12
2	7
3	3
4	2

An additional column can be included to show the total frequencies

Number of pets	Male students	Female students	Frequency
0	6	3	9
1	8	4	12
2	2	5	7
3	1	2	3
4	0	2	2
Total	17	16	33

An additional row shows the total for the number of male and female students

Apply 1

1 Use the following information to complete the tally chart.

3	2	2	1	3	4	0	1	3	0	2	1
1	4	3	2	2	1	3	2	3	1	1	0
4	3	2	2	0	1	0	1	1	0	2	3

Number	Tally	Frequency
0		
1		
2		
3		
4		

2 The tally chart shows the number of bedrooms in properties advertised in a newspaper.

Bedrooms	Tally	Frequency
1	ⵌⵌ ⵌⵌ	
2	ⵌⵌ ⵌⵌ IIII	
3	ⵌⵌ II	
4	IIII	
5	I	

a Copy and complete the tally chart including the frequency column.

b Use your table to answer the following.

   i How many properties were surveyed altogether?

   ii How many properties had three bedrooms?

   iii What is the modal number of bedrooms?

3 Peter is completing a tally chart for the following information.

5 7 6 8 8 6 5 7 8 6
6 5 5 7 5 9 5 7 6 5

His tally chart looks like this:

Number	Tally	Frequency
5	ⵌⵌ II	7
6	ⵌⵌ	5
7	ⵌⵌ	5
8	III	3
9	I	1
	Total	21

a How can you tell that Peter's tally chart is incorrect?
   Give a reason for your answer.

b Correct the tally chart.

4 The following information shows the marks obtained by a class in a test.

15 17 23 25 22 18 17 14 12 10 14
18 21 22 23 17 14 10 11 16 18 21
21 22 19 14 13 21

Use this information to complete the tally chart below.

Number	Tally	Frequency
10–13		
14–17		
18–21		
22–25		

5 The tally chart shows the heights (in centimetres to the nearest centimetre) of bushes in a garden centre.

The interval $10 < h \leqslant 15$ includes all heights from 10 cm to 15 cm but not including 10 cm

Number	Tally	Frequency
$5 < h \leqslant 10$	ЖЖ III	
$10 < h \leqslant 15$	ЖЖ ЖЖ	
$15 < h \leqslant 20$	ЖЖ II	
$20 < h \leqslant 25$	ЖЖ	

a Copy and complete the table and use it to answer the following.

  i How many bushes were surveyed in the garden centre?

  ii How many bushes had heights between 10 and 15 cm?

  iii How many bushes had heights above 15 cm?

b Adnan says that the modal height is 14 cm.
Is he correct?
Give a reason for your answer.

6 The two-way table shows the number of students who wear and do not wear glasses.

	Boys	Girls
Glasses	8	17
No glasses	15	24

Use the table to answer the following questions.

a How many people wear glasses?

b How many girls were in the survey?

c How many boys do not wear glasses?

7 50 students are asked if they are left-handed or right-handed.

	Boys	Girls
Left-handed	7	6
Right-handed	20	17

Use the table to answer the following questions.

a How many people are left-handed?

b How many boys were in the survey?

c What fraction of those asked are right-handed girls?

8 Students in a school were asked whether they had school dinners or packed lunches. Their results are shown in the table below.

	Boys	Girls
School dinner	24	16
Packed lunch	12	32

Use the table to answer the following questions.

a How many boys had a packed lunch?

b How many students had school dinners?

c How many girls were asked?

9 Students in a school were asked about their favourite pet.
Their results are shown in the table below.

	Dogs	Cats	Fish
Boys	24	16	48
Girls	12	24	38

Use the table to answer the following questions.

a How many boys like cats?

b How many students liked fish?

c How many girls were in the survey?

d How many students were asked altogether?

10 The table shows the different animals on a farm.

	Sheep	Cattle	Pigs
Male	80		90
Female		70	

Copy and complete the table if the farmer has:

- 130 sheep altogether

- 340 male animals

- 600 animals altogether.

11 In a school with 100 students, 52 boys played football, 18 girls played
football and 15 girls did not play football.

Copy and complete the following table.

	Played football	Did not play football
Boys		
Girls		

12

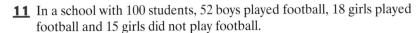

Prices for Costa Del Hot – per week

	7th April to 5th June	6th June to 21st July	22nd July to 5th Sept
Adult	£124	£168	£215
Child (5–16 years)	£89	£120	£199
Child (0–5 years)	Free	£12	£50

The Burridge family consists of 2 adults and 2 children aged 12 and 3 years.
The Gaudy family consists of 2 adults and 3 children aged 15, 11 and 7 years.

a How much will it cost the Burridge family for a one week
holiday from the 12th to the 19th June?

b How much will it cost the Gaudy family for a one week
holiday from the 7th to the 14th August?

Explore

◎ Collect some two-way tables of your own, for example, from a holiday brochure

◎ Write some questions making use of the tables and ask a friend to answer your questions

Investigate further

Learn 2 Collecting data

Examples:

a Categorise each of the following:
  i Weight of whales in an aquarium
  ii Favourite make of car of students in the sixth form
  iii Number of bottles of lemonade sold at a store each day
  iv Best pop group in the charts.

	Quantitative	Qualitative	Continuous	Discrete
i Weight of whales in an aquarium	✓		✓	
ii Favourite make of car of students in the sixth form		✓		
iii Number of bottles of lemonade sold at a store each day	✓			✓
iv Best pop group in the charts		✓		

b Write down one advantage and one disadvantage of the following methods of collecting data.
  i Personal surveys.
  ii Postal surveys.
  iii Direct observations.

  i This is the most common method of collecting data and involves an interviewer asking questions of the interviewee. This method is sometimes called a face-to-face interview.

  Advantages
  • The interviewer can ask more complex questions and explain them if necessary.
  • The interviewer is likely to be more consistent when they record the responses.
  • The interviewee is more likely to answer the questions than with postal or e-mail surveys.

Disadvantages
- This method of interviewing takes a lot of time and can be expensive.
- The interviewer can influence the answers and this may cause bias.
- The interviewee is more likely to lie or to refuse to answer a question.

The telephone survey is a special case of the personal survey with similar advantages and disadvantages

ii Postal surveys make use of mailing lists (or the electoral register) and involve people being selected and sent a questionnaire.

Advantages
- The interviewees can take their time answering and give more thought to the answer.
- The possibility of interviewer bias is avoided.
- The cost of a postal survey is usually lower.

Disadvantages
- Postal surveys suffer from low response rates which may cause bias.
- The process can take a long time to get questionnaires out and await their return.
- Different people might interpret questions in different ways when giving their answers.

The e-mail survey is a special case of the postal survey and an increasingly popular method for collecting data

iii Direct observation, as the name implies, means observing the situation directly. Direct observation might include, for example, counting cars at a motorway junction or observing someone to see what shopping they buy. It can take place over a short or a long period of time.

Advantages
- Direct observation is a reliable method which allows observations to take place in the interviewees 'own environment'.

Disadvantages
- For some experiments, the interviewee may react differently because they are being observed.
- This method takes a lot of time and can be expensive.

c Write down three requirements of a good questionnaire.

- **Appropriate** to the survey being carried out and not asking unnecessary questions.

For example, asking someone for their address may not be appropriate for most questionnaires ... unless it is a survey on where people live

- **Unbiased** so that they do not lead the respondent to give a particular answer.

For example, asking the question 'Do you believe we should have a new shopping centre?' is a biased (leading) question

- **Unambiguous** so that they are clear and straightforward to the respondent.

For example, asking the question 'Do you agree or disagree that we should have a new shopping centre?' is not very clear

Apply 2

1 For each of the following say whether the data is quantitative or qualitative.

   a The number of people at a cricket test match.

   b The weights of new born babies.

   c How many cars a garage sells.

   d Peoples' opinions of the latest Hollywood blockbuster.

   e The best dog at Crufts.

   f The time it takes to run the London Marathon.

   g The colour of baked beans.

   h How well your favourite football team played in their last match.

   i The number of text messages received in a day.

2 For each of the following say whether the data is discrete or continuous.

   a The number of votes for a party at a general election.

   b The exact ages of students in your class.

   c The number of beans in a tin.

   d The weight of rubbish each household produces each week.

   e How many people watch the 9 o'clock news.

   f How long it takes to walk to school.

   g The number of sheep Farmer Angus has.

   h The weights of Farmer Angus' sheep.

   i The heights of Year 10 students in your school.

3 Connect the following to their proper description.
   The first one has been done for you.

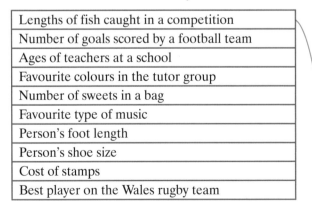

| Lengths of fish caught in a competition |
| Number of goals scored by a football team |
| Ages of teachers at a school |
| Favourite colours in the tutor group |
| Number of sweets in a bag |
| Favourite type of music |
| Person's foot length |
| Person's shoe size |
| Cost of stamps |
| Best player on the Wales rugby team |

Quantitative and discrete

Qualitative

Quantitative and continuous

4 In the following questionnaire identify whether the information requested is:

- quantitative and discrete.

- quantitative and continuous.

- qualitative.

Questionnaire

Age (years)	
Gender (M/F)	
Height (cm)	
Hand span (cm)	
Arm span (cm)	
Foot length (cm)	
Eye colour	
No. of brothers	
No. of sisters	
House number	
Favourite pet	

5 The following questions are taken from different surveys.
Write down one criticism of each question.
Rewrite the question in a more suitable form.

a How many hours of TV do you watch each week?

Less than 1 hour ☐   More than 1 hour ☐

b What is your favourite football team?

Real Madrid ☐   Luton Town ☐

c How do you spend your leisure time? (You can only tick one box.)

Doing homework ☐   Playing sport ☐   Reading ☐

Computer games ☐   On the Internet ☐   Sleeping ☐

d You do like football, don't you?

Yes ☐   No ☐

e How much do you earn each year?

Less than £10 000 ☐   £10 000 to £20 000 ☐   More than £20 000 ☐

f How often do you go to the cinema?

Rarely ☐   Sometimes ☐   Often ☐

g Do you or do you not travel by taxi?

Yes ☐   No ☐

6 Peter and Paul are writing questions for their coursework task.

   a This is one question from Peter's questionnaire:

> Skateboarding is an excellent pastime. Don't you agree?
> Tick one of the boxes.
>
> Strongly agree ☐   Agree ☐   Don't know ☐

   Write down two criticisms of Peter's question.

   b These questions are from Paul's questionnaire:

> Do you buy CDs? ☐ Yes   ☐ No
>
> If yes, how many CDs do you buy on average each month?
>
> ☐ 2 or less ☐ 3 or 4 ☐ 5 or 6 ☐ more than 6

   Write down two reasons why these are good questions.

7 Write down a definition for:

   a quantitative data

   b qualitative data

   c continuous data

   d discrete data.

8 Write down two advantages of undertaking a pilot survey.

9 Write down five things that make a good questionnaire.

Explore

Here are some investigations for you to consider.

◎ A company slogan states 'In tests 9 out of every 10 cats prefer it'

◎ Investigate ways in which the company could have arrived at this answer

◎ You might consider carrying out your own survey

◎ How might you get such results?

◎ In a supermarket poll of 100 people 85 said they had at least two cars!

◎ Is this correct?

◎ Investigate ways in which the company could have arrived at this answer

◎ How might you get such results?

Investigate further

Learn 3 Sampling methods

Examples:

You are given a list of 500 students (200 boys and 300 girls) and wish to choose a sample of 50.

Explain how you would use the following sampling methods.

a Convenience sampling
b Random sampling
c Systematic sampling
d Quota sampling
e Cluster sampling

a Convenience sampling or opportunity sampling means that you just take the first people who come along or those who are convenient to sample.

The likelihood is that you choose the first 50 students that you meet or otherwise choose 50 students from among your friends.

b Random numbers can be taken from random number tables or generated by a calculator using the $\boxed{\text{RAN}}$ or $\boxed{\text{RND}}$ button.

Assign each student a number between 0 and 499 and generate random numbers to choose 50 students.

c Systematic sampling is similar to random sampling except that you take every nth member of the population. The value n is found by dividing the population size by the sample size giving $\frac{500}{50} = 10$ so that every 10th student is chosen from the population.

A random number is used to start so that the number 4 would suggest taking the 4th, 14th, 24th, 34th... students.

d Quota sampling is popular in market research and involves choosing a sample with certain characteristics (the choice of who to ask is left to the interviewer).

The likely requirement is that you are asked to sample 20 boys and 30 girls.

e Cluster sampling is useful where the population is large and it is possible to split the population into smaller groups or clusters.

The most obvious choice is to consider tutor groups as clusters and sample the whole of two tutor groups, although this is not likely to result in exactly 50 students.

Apply 3

1 The following table shows the ordered ages for 100 people in a London shopping centre.

22	22	22	22	22	22	23	23	24	24
24	24	24	24	24	24	24	25	25	25
25	25	25	25	25	26	26	26	26	26
26	26	26	26	26	27	27	27	27	27
28	28	28	28	28	28	28	28	29	29
29	29	29	30	31	31	31	31	31	31
32	32	33	33	33	33	33	33	34	34
34	34	34	34	34	34	34	34	35	35
35	35	35	36	36	36	36	36	36	36
37	38	40	40	42	42	44	47	48	50

a Take a random sample of 20 people using the [RAN] or [RND] button on your calculator.
Find the mean of this sample.

b Take a systematic sample of 20 people using every fifth number.
Find the mean of this sample.

c Take a cluster sample of the first 20 people in the table.
Find the mean of this sample.

d Which sample do you think is most representative of the 100 people?
Give a reason for your answer.

2 Which sampling method is most appropriate for each of the following surveys? Give a reason for your answer.

a The amount of pocket money students receive in different year groups at your school.

b The favourite programmes of 15 to 19 year olds.

c The average life expectancy of people around the world.

d The favourite holiday destinations of people in the sixth form at your school.

e The opinion of two hundred 25 to 39 year olds on their favourite soap opera.

f Information on voting intentions at a general election.

3 Consider each of the following surveys and say whether the sample is representative. Give a reason for your answer.

   a Aiden is trying to find out what students in the school think about school dinners. He decides to use cluster sampling, asking the first 30 Year 7s as they leave the dining room.

   b Betty wants to find out the most popular names for newborn babies – she goes to the Internet, finds the relevant website and takes the first 100 names on the list.

   c Cameron wants to find out if all premiership footballers think that having overseas players is a good thing. He decides to take a 10% sample and ask all the players at Chelsea and Arsenal.

   d Davina wishes to check how many people travel on the underground. She telephones 100 people at home in the evening and asks them if they have travelled on the underground that week.

   e Eric undertakes a convenience sample of 20 friends to see if they wear glasses. He says that 45% of students at his college wear glasses.

4 Explain the difference between a random sample and a systematic sample.

5 Find two articles containing around 200 words each.
Find which article contains longer words.

   a Choose an appropriate sampling method and give a reason for your choice.

   b Choose an appropriate sample size and give a reason for your choice.

6 A college wishes to undertake a survey on the smoking habits of its students. Explain how you would take:

   a a random sample of 100 students

   b a systematic sample of 100 students.

7 Write down one advantage and one disadvantage of each of these sampling methods.

   a Convenience sampling

   b Random sampling

   c Systematic sampling

   d Quota sampling

Explore

   ◎ Design your own questionnaire and undertake a survey

   ◎ You should consider a hypothesis and write some suitable questions

   ◎ Design a questionnaire and undertake a pilot survey

   ◎ Amend your questionnaire and explain any improvements

   ◎ Consider a suitable sampling method – explain why you have chosen this method

   ◎ Carry out the survey and comment on your results

Investigate further

Collecting data

The following exercise tests your understanding of this chapter, with the questions appearing in order of increasing difficulty.

1 a Here is some information about the students in set 9U1.

	Boys	Girls
Walk to school	13	11
Do not walk to school	4	7

   i How many students are there in set 9U1?

  ii How many students are boys?

 iii How many students walk to school?

  iv How many students do not walk to school?

   v Are there more girls or boys in the set? How many more are there?

b Here is some information about GCSE results in a school.

	Mathematics	Science
Number of students	235	207
Number of students gaining grade A*	17	21

   i What was the total number of examinations sat?

  ii How many more students sat mathematics than science?

 iii What was the total number of A*s achieved?

  iv How many mathematics students did not achieve a grade A*?

2 For each of the following, state whether the data is quantitative or qualitative.

 a The heights of the people watching a tennis match.

 b The colours of cars in a car park.

 c The sweetness of different orange juices.

 d The lengths of pencils in a pencil case.

 e The numbers of students in different classes.

 f The numbers of leaves on different types of trees.

 g The ages of people at a night club.

 h The musical abilities of students in a class.

3 For each of the following, state whether the data is continuous or discrete.

 a The heights of the people watching a tennis match.

 b The times taken by athletes to complete a race.

 c The numbers of sweets in a sample of packets.

 d The lengths of pencils in a pencil case.

 e The numbers of students in different classes.

 f The numbers of leaves on different types of trees.

 g The shoe sizes of people at a party.

 h The ages of people at a night club.

4 a Criticise each of the following questionnaire questions.

    i How many hours of television have you watched in the last 2 months?

    ii Do you or do you not watch news programmes?

b Criticise each of the following questionnaire questions and suggest alternatives to find out the required information.

    i What do you think about our new improved fruit juice?

    ii How much do you earn?

    iii Do you or do you not agree with the new bypass?

    iv Would you prefer to sit in a non-smoking area?

    v How often do you have a shower?

5 a Explain why the each of the following may not produce a truly representative sample.

    i Selecting people at random outside a supermarket.

    ii Selecting every 10th name from an electoral register starting with M.

    iii Selecting names from a telephone directory.

b In a telephone poll conducted one morning, 20 people were asked whether they regularly used a bus to get to work. Give three reasons why this sample might not be truly representative.

6 a Explain how you could use the electoral roll to obtain a simple random sample.

b A college wishes to undertake a survey on the part-time employment of its students. Explain how you would take a random sample of 100 students.

Try a real past exam question to test your knowledge:

7 Nick asked his friends what their favourite colour was.
His results are shown in the table.

Colour	Tally	Frequency
Blue	ⵘ ⵘ IIII	
Red	IIII	
Green	ⵘ ⵘ	
Black	ⵘ III	

a i Complete the frequency column.

    ii How many friends did Nick ask altogether?

b Complete the pictogram to show Nick's results.

Use the symbol ⵣ to represent 4 friends.

Blue	
Red	
Green	
Black	

Spec B Modular, Module 1, Nov 02

17 Percentages

OBJECTIVES

F ▶ **Examiners would normally expect students who get an F grade to be able to:**

Understand that percentage means 'out of 100'

Change a percentage to a fraction or a decimal and vice versa

E ▶ **Examiners would normally expect students who get an E grade also to be able to:**

Compare percentages, fractions and decimals

Work out a percentage of a given quantity

Calculate simple interest

D ▶ **Examiners would normally expect students who get a D grade also to be able to:**

Increase or decrease a quantity by a given percentage

Express one quantity as a percentage of another

C ▶ **Examiners would normally expect students who get a C grade also to be able to:**

Work out a percentage increase or decrease

What you should already know ...

- Place values in decimals (for example, $0.6 = \frac{6}{10}$, $0.06 = \frac{6}{100}$)
- How to put decimals in order of size
- How to express fractions in their lowest terms (or simplest form)
- How to change a fraction to a decimal and vice versa

VOCABULARY

Percentage – a number of parts per hundred, for example, 15% means $\frac{15}{100}$

Numerator – the number on the top of a fraction

Numerator ⟶ $\frac{3}{8}$ ⟵ Denominator

Denominator – the number on the bottom of a fraction

Interest – money paid to you by a bank, building society or other financial institution if you put your money in an account or the money you pay for borrowing from a bank

Principal – the money put into the bank or borrowed from the bank

Rate – the percentage at which interest is added, usually expressed as per cent per annum (year)

Time – usually measured in years for the purpose of working out interest

Amount – the total you will have in the bank or the total you will owe the bank, at the end of the period of time

Balance – the amount of money you have in your bank account or the amount of money you owe after you have paid a deposit

Deposit – an amount of money you pay towards the cost of an item, with the rest of the cost to be paid later

Discount – a reduction in the price, perhaps for paying in cash or paying early

VAT (Value Added Tax) – a tax that has to be added on to the price of goods or services

Depreciation – a reduction in value, for example, due to age or condition

Credit – when you buy goods 'on credit' you do not pay all the cost at once; instead you make a number of payments at regular intervals, often once a month

Learn 1　Percentages, fractions and decimals

Examples:　　　**a** Convert the following decimals to percentages.

　　i 0.76
　　ii 0.7

i 0.76	$= 0.76 \times 100\%$	$= 76\%$
ii 0.7	$= 0.7 \times 100\%$	$= 70\%$

To change a decimal to a percentage, multiply by 100 (the digits will move 2 places to the left)

b Convert the following percentages to decimals.

　　i 16%
　　ii 4%

i 16%	$= 16\% \div 100$	$= 0.16$
ii 4%	$= 4\% \div 100$	$= 0.04$

To change a percentage to a decimal, divide by 100 (the digits will move 2 places to the right)

c Convert the following fractions to percentages.

　　i $\frac{27}{100}$
　　ii $\frac{9}{210}$

i $\frac{27}{100}$	$= \frac{27}{100} \times 100\%$	$= 27\%$
ii $\frac{9}{20}$	$= \frac{9}{20} \times 100\%$	$= 45\%$

To change a fraction to a percentage, multiply by 100

d Convert the following percentages to fractions.

　　i 20%
　　ii 85%

i 20%	$= \frac{20}{100} = \frac{20^{\,1}}{100_{\,5}}$	$= \frac{1}{5}$
ii 85%	$= \frac{85}{100} = \frac{85^{\,17}}{100_{\,20}}$	$= \frac{17}{20}$

To change a percentage to a fraction, put it over 100 and cancel to lowest terms

To compare fractions, decimals and percentages, change them all to percentages

Apply 1

1 Change each decimal to a percentage.

   a 0.27   **c** 0.08   **e** 1.28   **g** 0.125

   b 0.15   **d** 0.8   **f** 3.4   **h** $0.\dot{3}$

2 Change each percentage to a decimal.

   a 44%   **e** 105%

   b 59%   **f** 225%

   c 70%   **g** 37.5%

   d 3%   **h** 14.25%

3 Change each fraction to a percentage.

   a $\frac{57}{100}$ **c** $\frac{3}{10}$ **e** $\frac{1}{5}$ **g** $\frac{13}{20}$

   b $\frac{7}{100}$ **d** $\frac{9}{50}$ **f** $\frac{12}{25}$ **h** $\frac{3}{8}$

4 Change each percentage to a fraction, in its simplest form.

   a 99%   **e** 8%

   b 30%   **f** 75%

   c 26%   **g** $7\frac{1}{2}$%

   d 55%   **h** 12.5%

5 Place these in order of size, smallest first.

   a $\frac{2}{5}$, 0.36, 42%   **b** 28%, $\frac{1}{4}$, 0.3   **c** 0.1, 9%, $\frac{2}{25}$

6 Jake says that 34% is less than a third.
   Is he right?

7 Gina says that one eighth is 1.25%.
   What mistake has she made?

8 Change each of these fractions to a percentage.

   $\frac{3}{5}$, $\frac{9}{20}$, $2\frac{2}{5}$, $\frac{8}{25}$

   Use your answers to fill in the number square, as if they are the answers
   to crossword clues. (One answer has been entered in the grid for you.)

Learn 2 Working out a percentage of a given quantity

Examples: **a** Find 35% of £5.40

Without a calculator	Using a calculator
$10\% = £5.40 \div 10$ $= 54p$ $5\% = \frac{1}{2}$ of 10% $= 27p$	$1\% = £5.40 \div 100$ $35\% = £5.40 \div 100 \times 35$ $= £1.89$ or $\qquad 35\% = £0.35$ $0.35 \times £5.40 = £1.89$

10%	54p
+10%	+54p
+10%	+54p
+ 5%	+27p
35%	£1.89

b Calculate the simple interest on £300 for 2 years at 8%

Without a calculator	Using a calculator
8% of £100 = £8 8% of £300 = $3 \times £8 = £24$ For 2 years the interest is twice this : $2 \times £24 = £48$	$8\% = 8 \div 100$ $8 \div 100 \times £300 = £24$ For 2 years the interest is twice this : $2 \times £24 = £48$

$$\text{Formula: Simple interest} = \frac{\textbf{Principal} \times \textbf{Rate} \times \textbf{Time (in years)}}{\textbf{100}}$$

$$\textbf{SI} = \frac{\textbf{P} \times \textbf{R} \times \textbf{T}}{\textbf{100}}$$

Apply 2

Work out the following:

1 10% of 60 m

2 7% of £300

3 25% of £3.60

4 20% of 40 kg

5 80% of 40 kg

6 5% of £8.20

7 30% of 50 cm

8 15% of £70

9 75% of 9.6 m

10 $17\frac{1}{2}\%$ of £540

11 The simple interest on £400 for 2 years at 6% per annum

12 The simple interest on £200 for 7 years at 5% per annum

13 The simple interest on £1000 for 2 years at $4\frac{1}{2}$% per annum

14 The simple interest on £5000 for 3 years at 8% per annum

15 The final amount if £500 is invested for 4 years at simple interest of 3% per annum

 Work out the following:

16 35% of £2.40

17 52% of £850

18 40% of 19.5 m

19 16% of £94.80

20 18% of 175 g

21 6% of £4.50

22 84% of 35 kg

23 27% of 60 cm

24 130% of £35.58

25 $37\frac{1}{2}$% of £11.52

26 The simple interest on £950 for 6 years at 7% per annum

27 The simple interest on £6520 for 4 years at 3.8% per annum

28 The simple interest on £4965 for 5 years at $12\frac{1}{2}$% per annum

29 The simple interest on £14 600 for 3 years at $5\frac{3}{4}$% per annum

30 The total owed if £1575 is borrowed for 2 years at simple interest of 14.2% per annum

Learn 3 Increasing or decreasing by a given percentage

Examples:

a Parveen's bus fare to town is 80p. The bus fares go up by 5%. How much is the new fare?

10% of 80p = 8p
5% of 80p = 4p
New fare = 84p
	Original fare = 100%
	New fare = (100+5)%
	          =105% = 1.05
	New fare = 1.05×80p
	          = 84p

b Find the new price of a £350 TV after a 4% reduction.

1% of £350 = £3.50
4% of £350 = £3.50 × 4 = £14
New price = £350 – £14
          = £336
	Original price = 100%
	New price = (100−4) %
	          = 96% = 0.96
	New price = 0.96 × £350
	          = £336

Apply 3

 1 Increase 25 cm by 10%.

 2 Decrease 700 g by 5%.

 3 Decrease £450 by 20%.

 4 Increase £3 by 8%.

5 Get Real!
Todd is paid £300 per week.
He gets a 4% pay rise.
What is his new weekly pay?

6 Get Real!
A package holiday is priced at £660.
Gary gets a 10% discount for booking before the end of January.
How much does he pay?

7 Get Real!
Emma gets a 15% discount on purchases from Aqamart.
How much does she pay for a TV priced at £500?

8 Get Real!
Parveen's bus fare to town is 80p.
The bus fares go up by 5%.
How much will Parveen now have to pay to travel to town?

9 Increase 125 cm by 16%.

10 Increase 340 g by 5%.

11 Decrease £560 by 12%.

12 Decrease £9.50 by 22%.

13 Get Real!

The population of Baytown was 65 000 in 1990.
By the year 2000, Baytown's population had gone up by 27%.
What was the population in 2000?

14 Get Real!

Becky buys a new car for £12 000.
Over 2 years, it depreciates by 45%.
What is the value of the car after 2 years?

15 Get Real!

Liam sold 540 DVDs in June.
In July, his sales dropped by 15%.
How many DVDs did he sell in July?

16 Get Real!

The bill for a repair is £57.30
VAT at $17\frac{1}{2}$% has to be added to the bill.
What is the total cost of the repair?

Explore

◎ Kate wants to buy a music centre priced at £439

◎ She has to put down £100 deposit

◎ There are two ways she can pay the rest of the price (the balance)

1 The EasyPay option:
  • 10% credit charge on the balance
  • 6 equal monthly payments.

2 The PayLess option:
  • 3% added each month to the amount owing at the beginning of the month
  • pay £60 per month until the balance is paid off

(Note: in the last month Kate will only pay the remaining balance, not a full £60)

◎ Using a calculator, investigate these two options to advise Kate which one is best

◎ Would your advice be different if EasyPay charged 11% or PayLess charged $3\frac{1}{2}$% each month?

Investigate further

Learn 4 Expressing one quantity as a percentage of another

Examples:

a Express 27 g as a percentage of 300 g.

27 as a fraction of 300 is $\frac{27}{300}$

To convert a fraction to a percentage you multiply by 100:

$$\frac{{}^9 27}{{}_{100} 300} \times 100 = 9\%$$

Write them as a fraction and multiply by 100 to change to a percentage.

*Make sure both quantities are in the **same units**.*

b Express 84p as a percentage of £20.

Working in pounds, change 84p to £0.84

0.84 as a fraction of 20 is $\frac{0.84}{20}$

To convert a fraction to a percentage you multiply by 100:

$$\frac{0.84}{20} = \frac{0.84}{20} \times 100\%$$

$$= \frac{0.84}{20\ _1} \times 100^5\%$$

$$= 4.2\%$$

Working in pence, change £20 to 2000p

84 as a fraction of 2000 is $\frac{84}{2000}$

To convert a fraction to a percentage you multiply by 100:

$$\frac{84}{2000} = \frac{84}{2000} \times 100\%$$

$$= \frac{84\ ^{42}}{2000\ _{1000\ 10}} \times 100^1\%$$

$$= 4.2\%$$

Apply 4

 1 Express £22 as a percentage of £200.

 2 Express 12p as a percentage of 16p.

 3 Express 20 kg as a percentage of 50 kg.

 4 Express 21p as a percentage of £3.50.

 5 Express 85 mm as a percentage of 10 cm.

 6 Get Real!
There are 800 students in Uptown College.
96 of these students walk to college each day.
What percentage of the students walk to college?

 7 Get Real!
Chris has 50 books on his shelves.
29 of these books are science fiction.
What percentage of his books are science fiction?

 8 Get Real!
Kate has 6 girl cousins and 9 boy cousins.
What percentage of her cousins are boys?

 **9** Express £22.50 as a percentage of £450.

10 Express 7 cm as a percentage of 12.5 cm.

11 Express 65 g as a percentage of 5 kg.

12 Express £1190 as a percentage of £2800.

13 Express £17.64 as a percentage of £72.

14 Get Real!

Mel wanted to buy a sofa priced at £1450.
The salesman asked for a deposit of £348.
What percentage of the price was this?

15 Get Real!

Out of 3600 claims on household insurance,
522 were for broken windows.
What percentage of claims were for broken windows?

16 Get Real!

Grace has £6.25 in her purse.
She puts 20p in a charity box.
What percentage of her money has gone to charity?

Learn 5　Percentage increase and decrease

Examples:

a Find the percentage increase when the temperature goes up
from 20 °C to 26 °C.

Temperature increase = 6°
6 as a fraction of 20 is $\frac{6}{20}$
To convert a fraction to a percentage you multiply by 100:

$$\frac{6}{20} = \frac{6}{20} \times 100\%$$

$$= \frac{6}{20_1} \times 100^5\%$$

$$= 30\% \text{ increase}$$

> Make sure both quantities
> are in the **same units**

> Write this as a fraction and multiply
> by 100 to change to a percentage
> **(The original quantity has to be
> on the bottom of the fraction)**

b Find the percentage decrease when the price of a toy falls
from £12.50 to £11.75

Price decrease = £0.75
£0.75 as a fraction of £12.50 is $\frac{0.75}{12.50}$
To convert a fraction to a percentage you multiply by 100:

$$\frac{0.75}{12.50} = \frac{0.75}{12.50} \times 100\%$$

$$= \frac{0.75}{12.50_{25_1}} \times 100^{200^8}\%$$

$$= 6\% \text{ decrease}$$

c Find the percentage profit when an object is bought for £200 and sold for £256.

Price increase = £56

£56 as a fraction of £200 is $\frac{56}{200}$

To convert a fraction to a percentage you multiply by 100:

$$\frac{56}{200} = \frac{56}{200} \times 100\%$$

$$= \frac{56}{200_2} \times 100^{10}\%$$

$$= 28\% \text{ increase}$$

Apply 5

1 The price of a packet of biscuits goes up from 30p to 36p.
Find the percentage increase.

2 The price of a computer drops from £250 to £225.
Find the percentage decrease.

3 The population of a village goes up from 400 to 436.
Find the percentage increase.

4 Becky's curtains were 60 cm long before she washed them.
After the wash they were only 51 cm long.
Find the percentage decrease in length.

5 Sam buys a guitar for £125 and sells it for £160.
Find his percentage profit.

6 A landlord puts the rent on a flat up from £280 per month
to £301 per month.
Find the percentage increase.

7 A music system was priced at £240.
In the sale, the price dropped to £186.
Find the percentage decrease.

8 The price of a house goes down from £166 000 to £141 100.
Find the percentage decrease.

9 Callum buys a car for £2450 and does some repairs.
He sells the car for £2989.
Find his percentage profit.

10 Zoe says the answer to question 9 is 18%.
What mistake has she made?

Percentages

The following exercise tests your understanding of this chapter, with the questions appearing in order of increasing difficulty.

1 Copy and complete the table below.
Write each fraction in its simplest form.

Decimal	Fraction	Percentage
0.3		
0.55		
0.875		
	$\frac{1}{3}$	
	$\frac{3}{4}$	
	$\frac{4}{5}$	
		1%
		$2\frac{1}{2}$%
		48%
1.25		
	$2\frac{2}{3}$	
		320%

2 a Guy Fawkes bought a box of 75 matches to light his bonfire. The wood was damp and it took him 16% of his matches to light the fire. How many matches did he use?

b A DJ has 400 CDs, 65% of which are dance music.
How many dance music CDs does she have?

c A cricket team scored 256 runs altogether. Their leading batsman scored $37\frac{1}{2}$% of the runs. How many runs did he score?

d A shop assistant is paid 8% of all furniture sales as his monthly bonus.
What bonus did he get on sales of £4600?

3 a Write 60p as a percentage of £1.20

b What is 150 m as a percentage of 3 km?

c David has bought Victoria an 18 carat gold bracelet.
Pure gold is 24 carat.
What percentage of Victoria's bracelet is gold?

d A bag of sand is labelled as 50 kg.
It actually contains 2.5% more.
How much sand does it contain?

e Ms Berry has picked 1.2 kg of blackberries for making jam.
She needs 15% more to make her recipe.
What weight of blackberries does the recipe require?

f The amount of evaporation while whisky is maturing in the vat is known in the trade as the 'Angel's Share'. A vat of whisky originally held 55 litres. Evaporation reduced this by 16%. How many litres were left in the vat after the angels had received their share?

4 a A book is designed to have 650 pages.
When the author finished the manuscript he found he had written 754 pages.
What percentage increase is this?

b A box of 144 pens is bought for £10 and the individual pens are sold at 10p each. What is the percentage profit?

c Toad of Toad Hall bought his latest car for £18 000. A week later he crashed it and, after repair, sold it for £11 700. What was his percentage loss?

d 100 apples are bought for £17 but 5% are found to be damaged and not saleable. The rest are sold at 20p each. What is the percentage profit?

e Pythagoras makes a calculator error while using his famous theorem!
He wants to find the value of $\sqrt{112}$ but instead he finds $\sqrt{121}$.
What is the percentage error in his calculation?

Try a real past exam question to test your knowledge:

5 a A year ago, Mark was 185 cm tall.
He is now 4.3% taller.
How tall is Mark now?

b A year ago Vicky weighed 151 lb.
She now weighs 164 lb.
Calculate the percentage increase in Vicky's weight.

Spec B Modular, Module 3, Nov 02

Glossary

Acute angle – an angle between 0° and 90°

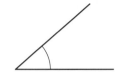

Algebraic expression – a collection of terms separated by + and – signs such as $x + 2y$ or $a^2 + 2ab + b^2$

Alternate angles – the angles marked a, which appear on opposite sides of the transversal

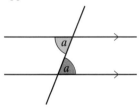

Amount – the total you will have in the bank or the total you will owe the bank, at the end of the period of time

Area – the amount of enclosed space inside a shape

Average – a single value that is used to represent a set of data

Axis (pl. **axes**) – the lines used to locate a point in the coordinates system; in two dimensions, the x-axis is horizontal, and the y-axis is vertical. This system of Cartesian coordinates was devised by the French mathematician and philosopher René Descartes

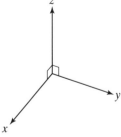

In three dimensions, the x- and y-axes are horizontal and at right angles to ech other and the z-axis is vertical

Back-to-back stem-and-leaf diagram – a stem-and-leaf diagram used to represent two sets of data

Number of minutes to complete a task

Leaf (units) Girls	Stem (tens)	Leaf (units) Boys
7 7 6 5 4 2 2	1	1 6 7 8 9
7 6 4 3 2 1	2	2 2 7 7 7 8 9
7 0	3	1 4 6

Key 3 | 2 represents 23 minutes

Key : 3 | 4 represents 34 minutes

Balance – the amount of money you have in your bank account or the amount of money you owe after you have paid a deposit

Bar chart – in a bar chart, the frequency is shown by the height (or length) of the bars. Bar charts can be vertical or horizontal

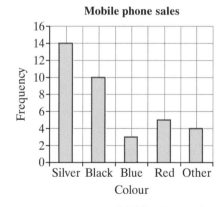

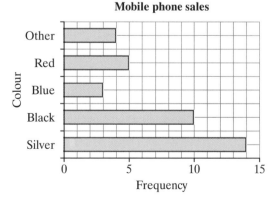

Bearing – an angle measured clockwise from North; all bearings should be written as three figure numbers, for example, 125° or 045°

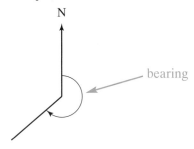

Categorical data – see qualitative data

Chord – a straight line joining two points on the circumference of a circle

Circle – a shape formed by a set of points that are always the same distance from a fixed point (the centre of the circle)

Circumference – the perimeter of a circle

Cluster sampling – this is useful where the population is large and it is possible to split the population into smaller groups or clusters

Collect like terms – to group together terms of the same variable, for example, $2x + 4x + 3y = 6x + 3y$

Common factor – factors that are in common for two or more numbers, for example,

the factors of 6 are 1, 2, 3, 6
the factors of 9 are 1, 3, 9
the common factors are 1 and 3

Common fraction – see fraction

Concave polygon – a polygon with at least one interior reflex angle

Congruent – exactly the same size and shape; one of the shapes might be rotated or flipped over

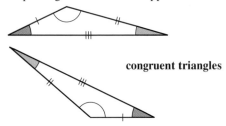

congruent triangles

Continuous data – data that can be measured and take any value; length, weight and temperature are all examples of continuous data

Convenience or **opportunity sampling** – a survey that is conducted using the first people who come along, or those who are convenient to sample (such as friends and family)

Convex polygon – a polygon with no interior reflex angles

Coordinates – a system used to identify a point; an x-coordinate and a y-coordinate give the horizontal and vertical positions

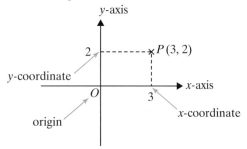

Correlation – a measure of the relationship between two sets of data; correlation is measured in terms of type and strength

Strength of correlation

The strength of correlation is an indication of how close the points lie to a straight line (perfect correlation)

Strong correlation   **Weak correlation**

 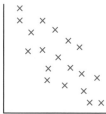

Correlation is usually described in terms of strong correlation, weak correlation or no correlation

Type of correlation

Positive correlation   **Negative correlation**

In positive correlation an increase in one set of variables results in an increase in the other set of variables

In negative correlation an increase in one set of variables results in a decrease in the other set of variables

Zero or no correlation

Zero or no correlation is where there is no obvious relationship between the two sets of data

Corresponding angles – the angles marked c, which appear on the same side of the transversal

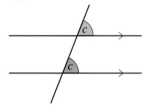

Counting number or **natural number** – a positive whole number, for example, 1, 2, 3, ...

Credit – when you buy goods 'on credit' you do not pay all the cost at once; instead you make a number of payments at regular intervals, often once a month

Cube number – a cube number is the outcome when a number is multiplied by itself then multiplied by itself again; cube numbers are 1, 8, 27, 64, 125, ...

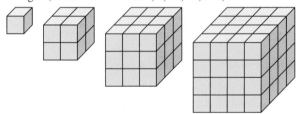

The rule for the nth term of the cube numbers is n^3

Cube root – the cube root of a number such as 125 is a number whose outcome is 125 when multiplied by itself then multiplied by itself again

Data – information that has been collected

Data collection sheets – these are used to record the responses to the different questions on a questionnaire; they can also be used with computers to load data onto a database

Decagon – a polygon with ten sides

Decimal – a number in which a decimal point separates the whole number part from the decimal part, for example, 24.8

Decimal fraction – a fraction consisting of tenths, hundredths, thousandths, and so on, expressed in a decimal form, for example, 0.65 (6 tenths and 5 hundredths)

Decimal places – the digits to the right of a decimal point in a number, for example, in the number 23.657, the number 6 is the first decimal place (worth $\frac{6}{10}$), the number 5 is the second decimal place (worth $\frac{5}{100}$) and 7 is the third decimal place (worth $\frac{7}{1000}$); the number 23.657 has 3 decimal places

Denominator – the number on the bottom of a fraction

Deposit – an amount of money you pay towards the cost of an item, with the rest of the cost to be paid later

Depreciation – a reduction in value, for example, due to age or condition

Diameter – a chord passing through the centre of a circle; the diameter is twice the length of the radius

Digit – any of the numerals from 0 to 9

Dimension – the measurement between two points on the edge of a shape

Directed number – a number with a positive or negative sign attached to it; it is often seen as a temperature, for example, -1, $+1$, $+5$, -3°C, $+2$°C, ...

Direct observation – collecting data first-hand, for example, counting cars at a motorway junction or observing someone shopping

Discount – a reduction in the price, perhaps for paying in cash or paying early

Discrete data – data that can only be counted and take certain values, for example, numbers of cars (you can have 3 cars or 4 cars but nothing in between, so $3\frac{1}{2}$ cars is not possible)

Equidistant – the same distance; if A is equidistant from B and C, then AB and AC are the same length

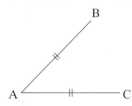

Equilateral triangle – a triangle with 3 equal sides and 3 equal angles – each angle is 60°

Equivalent fraction – a fraction that has the same value as another, for example, $\frac{3}{5}$ is equivalent to $\frac{30}{50}$, $\frac{6}{10}$, $\frac{60}{100}$, $\frac{15}{25}$, $\frac{1.5}{2.5}$, ...

Estimate – find an approximate value of a calculation; this is usually found by rounding all of the numbers to one significant figure, for example, $\frac{20.4 \times 4.3}{5.2}$ is approximately $\frac{20 \times 4}{5}$ where each number is rounded to 1 s.f., the answer can be worked out in your head to give 16

Expand – to remove brackets to create an equivalent expression (expanding is the opposite of factorising)

Exponent – see index

Exterior angle – the angle between one side of a polygon and the extension of the adjacent side

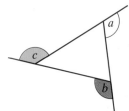

a, b and c are exterior angles

Factor – a natural number which divides exactly into another number (no remainder); for example, the factors of 12 are 1, 2, 3, 4, 6, 12

Factorise – to include brackets by taking common factors (factorising is the opposite of expanding)

Fibonacci numbers – a sequence where each term is found by adding together the two previous terms

1,   1,   2,   3,   5,   8,   13,   21,   ...

1+1 1+2 2+3 3+5 5+8 8+13

Fraction or **simple fraction** or **common fraction** or **vulgar fraction** – a number written as one whole number over another, for example, $\frac{3}{8}$ (three eighths), which has the same value as $3 \div 8$

Frequency diagram – a frequency diagram is similar to a bar chart except that it is used for continuous data. In this case, there are usually no gaps between the bars

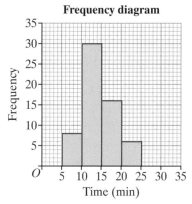

Frequency diagram

Frequency table or **frequency distribution** – a table showing how frequently each quantity occurs, for example,

Number in family	2	3	4	5	6	7	8
Frequency	2	3	8	4	2	0	1

Gradient – a measure of how steep a line is

$$\text{Gradient} = \frac{\text{change in vertical distance}}{\text{change in horizontal distance}}$$
$$= \frac{y}{x}$$

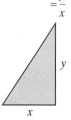

Greater than ($>$) – the number on the left-hand side of the sign is larger than that on the right-hand side

Grouped data – data that has been grouped into specific intervals

Heptagon – a polygon with seven sides

Hexagon – a polygon with six sides

Highest common factor (HCF) – the highest factor that two or more numbers have in common, for example,

the factors of 16 are 1, 2, 4, 8, 16
the factors of 24 are 1, 2, 3, 4, 6, 8, 12, 24
the common factors are 1, 2, 4, 8
the highest common factor is 8

Horizontal – from left to right; parallel to the horizon

Horizontal

Improper fraction or **top-heavy fraction** – a fraction in which the numerator is bigger than the denominator, for example, $\frac{13}{5}$, which is equal to the mixed number $2\frac{3}{5}$

Index or **power** or **exponent** – the index tells you how many times the base number is to be multiplied by itself

So $5^3 = 5 \times 5 \times 5$

Index notation – when a product such as $2 \times 2 \times 2 \times 2$ is written as 2^4, the number 4 is the index (plural **indices**)

Indices – the plural of index

Integer – any positive or negative whole number or zero, for example, $-2, -1, 0, 1, 2, ...$

Intercept – the y-coordinate of the point at which the line crosses the y-axis

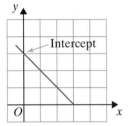

Interest – money paid to you by a bank, building society or other financial institution if you put your money in an account or the money you pay for borrowing from a bank

Interior angle – an angle inside a polygon

a, b, c, d and e are interior angles

Irregular polygon – a polygon whose sides and angles are not all equal (they do not all have to be different)

Isosceles trapezium – a quadrilateral with one pair of parallel sides. Non-parallel sides are equal

Isosceles triangle – a triangle with 2 equal sides and 2 equal angles; the equal angles are called **base angles**

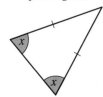

the x angles are base angles

Kite – a quadrilateral with two pairs of equal adjacent sides

Least common multiple (LCM) – the least multiple which is common to two or more numbers, for example,

the multiples of 3 are 3, 6, 9, 12, 15, 18, 24, 27, 30, 33, 36, ...
the multiples of 4 are 4, 8, 12, 16, 20, 24, 28, 32, 36, ...
the common multiples are 12, 24, 36, ...
the least common multiple is 12

Less than ($<$) – the number on the left-hand side of the sign is smaller than that on the right-hand side

Linear expression – a combination of terms where the highest power of the variable is 1

Linear expressions	Non-linear expressions
x	x^2
$x + 2$	$\frac{1}{x}$
$3x + 2$	$3x^2 + 2$
$3x + 4y$	$(x + 1)(x + 2)$
$2a + 3b + 4c + ...$	x^3

Line graph – a line graph is a series of points joined with straight lines

Temperature graph

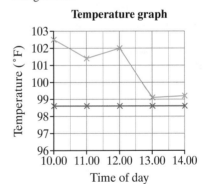

Time of day

Line of best fit – a line drawn to represent the relationship between two sets of data. Ideally it should only be drawn where the correlation is strong, for example,

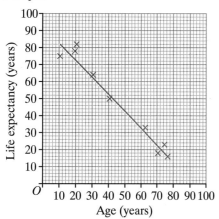

Line segment – the part of a line joining two points

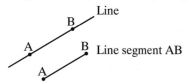

Lower bound – this is the minimum possible value of a measurement, for example, if a length is measured as 37 cm correct to the nearest centimetre, the lower bound of the length is 36.5 cm

Mean – found by calculating $\dfrac{\text{the total of all the values}}{\text{the number of values}}$

Median – the middle value when all the values have been arranged in order of size; for an even set of numbers, the median is the mean of the two middle values

Midpoint – the middle point of a line

Mixed number or **mixed fraction** – a number made up of a whole number and a fraction, for example, $2\frac{3}{5}$, which is equal to the improper fraction $\frac{13}{5}$

Modal class – the class with the highest frequency

Mode – the value that occurs most often

Multiple – the multiples of a number are the products of the multiplication tables, for example, the multiples of 3 are 3, 6, 9, 12, 15, ...

Natural number – see counting number

Negative number – a number less than 0; it is written with a negative sign, for example, – 1, – 3, – 7, – 11, ...

Nonagon – a polygon with nine sides

***n*th term** – this phrase is often used to describe a 'general' term in a sequence; if you are given the *n*th term, you can use this to find the terms of a sequence

Numerator – the number on the top of a fraction

Obtuse angle – an angle greater than 90° but less than 180°

Octagon – a polygon with eight sides

Opportunity sampling – see convenience sampling

Ordered stem-and-leaf diagram – a stem-and-leaf diagram where the data is placed in order

Number of minutes to complete a task

Stem (tens)	Leaf (units)
1	1 6 7 8 9
2	2 2 7 7 7 8 9
3	1 4 6

Key : 3 | 4 represents 34 minutes

Origin – the point (0, 0) on a coordinate grid

Outlier – a value that does not fit the general trend, for example,

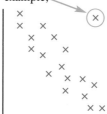

Parallel lines – two lines that never meet and are always the same distance apart

Parallelogram – a quadrilateral with opposite sides equal and parallel

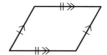

Pentagon – a polygon with five sides

Percentage – a number of parts per hundred, for example, 15% means $\frac{15}{100}$

Perimeter – the distance around an enclosed shape

Perpendicular lines – two lines at right angles to each other

Pictogram – in a pictogram, the frequency is shown by a number of identical pictures

Mobile phone sales

Colour	Frequency
Silver	
Black	
Blue	
Red	
Other	

Key 2 mobiles

Pie chart – in a pie chart, frequency is shown by the angles (or areas) of the sectors of a circle

Mobile phone sales

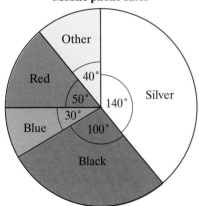

Pilot survey – a small-scale survey to check for any unforeseen problems with the main survey

Polygon – a closed two-dimensional shape made from straight lines

Positive number – a number greater than 0; it can be written with or without a positive sign, for example, 1, $+4$, 8, 9, $+10$, ...

Power – see index

Primary data – data that you collect yourself; this is new data and is usually collected for the purpose of a task or project (including GCSE coursework)

Prime number – a natural number with exactly two factors, for example, 2 (factors are 1 and 2), 3 (factors are 1 and 3), 5 (factors are 1 and 5), 7, 11, 13, 17, 23, ..., 59, ...

Principal – the money put into the bank or borrowed from the bank

Product – the result of multiplying together two (or more) numbers, variables, terms or expressions

Proper fraction – a fraction in which the numerator is smaller than the denominator, for example, $\frac{5}{13}$

Quadrant – one of the four regions formed by the x- and y-axes in the Cartesian coordinate system

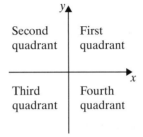

Quadrant (of a circle) – one quarter of a circle

Quadrilateral – a polygon with four sides

Qualitative or **categorical data** – data that cannot be measured using numbers, for example, type of pet, car colour, taste, peoples' opinions/feelings etc.

Quantitative data – data that can be counted or measured using numbers, for example, number of pets, height, weight, temperature, age, shoe size etc.

Quota sampling – this method involves choosing a sample with certain characteristics, for example, select 20 adult men, 20 adult women, 10 teenage girls and 10 teenage boys to conduct a survey about shopping habits

Radius – the distance from the centre of a circle to any point on the circumference

Random sampling – this requires each member of the population to be assigned a number; the numbers are then chosen at random

Range – a measure of spread found by calculating the difference between the largest and smallest values in the data, for example, the range of 1, 2, 3, 4, 5 is $5 - 1 = 4$

Rate – the percentage at which interest is added, usually expressed as per cent per annum (year)

Reciprocal – any number multiplied by its reciprocal equals one; one divided by a number will give its reciprocal, for example,

the reciprocal of 3 is $\frac{1}{3}$ because $3 \times \frac{1}{3} = 1$

Rectangle – a quadrilateral with four right angles, and opposite sides equal in length

Recurring decimal – a decimal with a repeating digit or group of digits, for example, 0.33333333333 ... (written as 0.3̇) or 0.25678678678678 ... (written as 0.25̇67̇8̇)

Reflex angle – an angle greater than 180° but less than 360°

Regular polygon – a polygon with all sides and all angles equal

Respondent – the person who answers the questionnaire

Revolution – one revolution is the same as a full turn or 360°

Rhombus – a quadrilateral with four equal sides and opposite sides parallel

Right angle – an angle of 90°

Right-angled triangle – a triangle with one angle of 90°

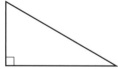

Round – give an approximate value of a number; numbers can be rounded to the nearest 1000, nearest 100, nearest 10, nearest integer, significant figures, decimal places, ... etc.

Scatter graph – a graph used to show the relationship between two sets of variables, for example, temperature and ice cream sales

Temperature against ice cream sales

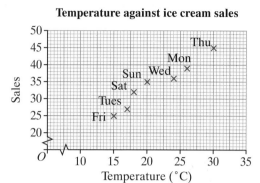

Secondary data – data that someone else has collected; this might include data in books, newspapers, magazines, etc. or data that has been loaded onto a database

Semicircle – one half of a circle

Sequence – a list of numbers or diagrams that are connected in some way

In this sequence of diagrams, the number of squares is increased by one each time:

The dots are included to show that the sequence continues

Shape – an enclosed space

Significant figures – the digits in a number; the closer a digit is to the beginning of a number then the more important or significant it is; for example, in the number 23.657, 2 is the most significant digit and is worth 20, 7 is the least significant digit and is worth $\frac{7}{1000}$; the number 23.657 has 5 significant digits

Simple fraction – see fraction

Simplify – to make simpler by collecting like terms

Simplify a fraction or **express a fraction in its simplest form** – to change a fraction to the simplest equivalent fraction; to do this divide the numerator and the denominator by a common factor (this process is called cancelling or reducing or simplifying the fraction)

Square – a quadrilateral with four equal sides and four right angles

Square number – a square number is the outcome when a number is multiplied by itself; square numbers are 1, 4, 9, 16, 25, ...

The rule for the nth term of the square numbers is n^2

Square root – a square root of a number such as 16 is a number whose outcome is 16 when multiplied by itself

Stem-and-leaf diagram – a way of arranging data using a key to explain the 'stem' and 'leaf' so that 3 | 4 represents 34

Number of minutes to complete a task

Stem (tens)	Leaf (units)
1	6 8 1 9 7
2	7 8 2 7 7 2 9
3	4 1 6

Key : 3 | 4 represents 34 minutes

Sum – to find the sum of two numbers, you add them together

Survey – a way of collecting data; there are a variety of ways of doing this, including face-to-face, or via telephone, e-mail or post using questionnaires

Systematic sampling – this is similar to random sampling except that it involves every nth member of the population; the number n is chosen by dividing the population size by the sample size

Tally chart – a useful way to organise raw data; the chart can be used to answer questions about the data, for example,

Number of pets	Tally
0	ⅢⅢ ⅢⅠ
1	ⅢⅢ ⅢⅢ ⅠⅠ
2	ⅢⅢ ⅠⅠ
3	ⅢⅠ
4	ⅠⅠ

The tallies are grouped into five so that

ⅠⅠⅠⅠ = 4
ⅢⅢ = 5
ⅢⅢⅠ = 6

This makes the tallies easier to read

Term – a number, variable or the product of a number and a variable(s) such as 3, x or $3x$

Terminating decimal – a decimal that ends, for example, 0.3, 0.33 or 0.3333

Tessellation – a pattern where one or more shapes are fitted together repeatedly leaving no gaps

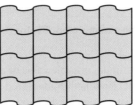

Time – usually measured in years for the purpose of working out interest

Time series – a graph of data recorded at regular intervals

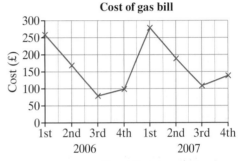

Cost of gas bill

Year/ Quarter

Top-heavy fraction – see improper fraction

Transversal – a line drawn across parallel lines

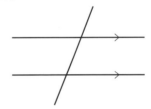

Trapezium (pl. trapezia) – a quadrilateral with one pair of parallel sides

Triangle – a polygon with three sides

Triangle numbers – 1, 3, 6, 10, 15, ...

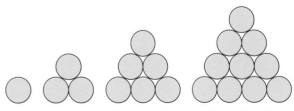

The rule for the nth term of the triangle numbers is $\frac{1}{2}n(n + 1)$

Two-way table – a combination of two sets of data presented in a table form, for example,

	Men	Women
Left-handed	7	6
Right-handed	20	17

Unit fraction – a fraction with a numerator of 1, for example, $\frac{1}{5}$

Upper bound – this is the maximum possible value of a measurement, for example, if a length is measured as 37 cm correct to the nearest centimetre, the upper bound of the length is 37.5 cm

Variable – a symbol representing a quantity that can take different values such as x, y or z

VAT (Value Added Tax) – a tax that has to be added on to the price of goods or services

Vertical – directly up and down; perpendicular to the horizontal

Vertical

Vulgar fraction – see fraction